# Parasitology Research Monographs

Volume 14

**Series Editor**

Heinz Mehlhorn, Department of Parasitology, Heinrich Heine University, Düsseldorf, Germany

This book series "Parasitology Research Monographs" presents carefully refereed volumes on selected parasitological topics. Parasites have an increasing impact on animal and human health in the present times of globalization and global warming. Parasites may be agents of diseases and- often at the same time- vectors of other agents of disease such as viruses, bacteria, fungi, protozoa and/or worms. The growth in knowledge of parasitic physiology, cell structure, biotechnological and genetic approaches, ecology, therapeutic capabilities, vaccination, immunology, diagnosis, transmission pathways and many other aspects of parasitology is increasing dramatically, even in the face of the breakthroughs that have already been made. Reflecting these most recent achievements and the importance of parasites as a threat to human and animal health, the series' broad scope concentrates on particularly hot topics that were recently covered by review articles in the journal "Parasitology Research" or in other journals. These reviews offer compact but intense insights into the ongoing research and into the methods and technologies used to control parasites. The volumes in the series build on these topics, and the volume editors are well-known experts in their respective fields. Each volume offers 10 to 20 comprehensive reviews covering all relevant aspects of the topic in focus.

More information about this series at http://www.springer.com/series/8816

Trevor N. Petney • Weerachai Saijuntha •
Heinz Mehlhorn
Editors

# Biodiversity of Southeast Asian Parasites and Vectors causing Human Disease

Springer

*Editors*
Trevor N. Petney
Departments of Zoology and Paleontology and Evolution
State Museum of Natural History
Karlsruhe, Germany

Cholangiocarcinoma Research Institute
Khon Kaen University
Khon Kaen, Thailand

Heinz Mehlhorn
Department of Parasitology
Heinrich Heine University
Düsseldorf, Germany

Weerachai Saijuntha
Walai Rukhavej Botanical Research Institute, Biodiversity and Conservation Research Unit
Mahasarakham University
Maha Sarakham, Thailand

ISSN 2192-3671 ISSN 2192-368X (electronic)
Parasitology Research Monographs
ISBN 978-3-030-71160-3 ISBN 978-3-030-71161-0 (eBook)
https://doi.org/10.1007/978-3-030-71161-0



This Springer imprint is published by the registered company Springer Nature Switzerland AG.
The registered company address is: Gewerbestrasse 11, 6330 Cham, Switzerland

# Preface

Southeast Asia is one of the most rapidly developing areas in the world with all of the corresponding changes to the environment: population growth, urbanization, landuse including the intensification of agriculture, ecosystem destruction including deforestation, hydrological changes, and soil degradation, pollution, as well as climate change. These are superimposed on a background of exceptionally high natural biodiversity. In this book, our aim is to examine how the changing environment is likely to influence the diversity of parasites present, with particular emphasis on those affecting the human population.

Chapter 1 discusses the situation facing populations of vertebrate and invertebrate host species in Southeast Asia. For wild populations, environmental changes including hunting for products for traditional medicines, the pet trade, and bush meat, as well as massive habitat destruction are likely to cause population declines and local extinctions. In contrast, synanthropic species are likely to benefit from human expansion.

In Chap. 2, Kittipong Chaisiri and Serge Morand use modern statistical methods to determine how the biodiversity and species association patterns of helminths change in relation to human landuse habitats using the *Rattus rattus*-complex from mainland Southeast Asia as the research object. They show that human landuse alteration in the form of peridomestic habitats leads to high levels of helminth diversity associated with the hosts' synanthropic behavior. In addition, they report the predominance of positive associations between parasite species compared with negative associations suggesting that affected areas may show multiple parasite transmission.

Chapter 3 addresses the topic of the importance of intestinal protozoa for humans throughout Southeast Asia. Kotchaphon Vaisusuk and Weerachai Saijuntha review the available information on *Giardia duodenalis*, *Entamoeba histolytica*, *Cryptosporidium*, and *Blastocystis* sp., providing tables with comprehensive data on the countries involved, the source population for the data, the prevalence found with the sample size, followed by the relevant references. In addition, new life cycle diagrams have been provided. The authors conclude that a relatively high prevalence of intestinal protozoa infection is common among children and HIV/AIDS patients in Southeast Asia, and that these intestinal protozoa should be considered of major public health importance.

Trematodes are frequent parasites of humans and animals throughout Southeast Asia affecting millions of people and causing considerable morbidity and mortality. Chapter 4 discusses this group of parasites and the freshwater snail intermediate hosts. There are a considerable number of human trematodes of public health importance. These can be found in the intestines, bile ducts of the liver, lungs, and blood vessels. The intermediate hosts include fish, snails as well as other invertebrates. Infection may occur by consumption of certain plant species living in association with freshwater habitats. Saijuntha et al. also consider the transmission route via contaminated food and water as well as a variety of traditionally cooked dishes.

In Chap. 5, Sanpool and co-workers consider the community of nematodes found in the human gut. They point out that our knowledge of the nematodes inhabiting the human gut is still far from complete. This chapter reviews and highlights important aspects of human nematode infections that affect public health in Southeast Asia. The authors emphasize the importance of molecular identification and genetic diversity providing the relevant phylogenetic trees for *Strongyloides stercoralis* and *S. fuelleborni*, hookworm, *Ascaris suum*, *Trichuris trichiura* and *T. suum* and *Enterobius vermicularis*.

Thanchomnang et al. in Chap. 6 review the medically important tapeworms that are of public health concern in Southeast Asia (*Taenia* spp., *Echinococcus* spp., *Hymenolepis* spp., *Spirometra* spp., *Sparganum proliferum*, *Dibothriocephalus* spp.) with emphasis on molecular identification and genetic diversity. For each species or species group, information is provided on classification, geographic distribution, the parasite's biology and infection, molecular identification, and genetic diversity.

Chapter 7 deals with blackflies and the parasites that they transmit. Black flies are vectors of human and livestock disease agents. Even without transmission of pathogens, black fly biting can affect human and animal welfare through nuisance and irritation. A total of 456 black fly species arranged in eight subgenera of the genus *Simulium* are found in Southeast Asia. A minimum of seven species will bite humans, two of which are considered as pests in northern Thailand. Three species have been found carrying filarial parasites of animals including two unidentified species of the genus *Onchocerca*. As most work on blackflies and the diseases that they transmit in their region have taken place only within the last two decades, further studies are required to evaluate the impact on human and animal health.

Ticks are the most important vectors of animal diseases and the second most important vector of human diseases worldwide. Nevertheless, in spite of a recent influx of publications, comparatively little is known about the role of ticks in disease transmission in Southeast Asia. In this region, they are found on most reptiles, birds, and mammals species, including humans. Some information already indicates that they act as vectors of viral, bacterial, and protozoan pathogens of animals and humans throughout the region. In Chap. 8, Saijuntha et al. provide an overview of the tick fauna of mainland and insular Southeast Asia, show the pathogens that the ticks transmit, and consider future changes in the structure of tick communities.

As with other areas of the world, global climate change is already impacting Southeast Asia. Petney and Andrews discuss these effects as well as potential future scenarios in Chap. 9. In Southeast Asia, substantial increases in temperature and regional changes in rainfall patterns are expected, with some areas experiencing weather conditions and others increased drought. Among the factors discussed are severe weather events, sea level rise, salinization, and drought induced wild fires. This information is then extrapolated to include parasite communities and the host communities on which they are reliant to complete their life cycles. As the authors indicate, potentially dramatic and dynamic changes will occur, however, our data base is still limited and studies of local and regional parasite communities are essential for our understanding of climate impact on human and animal health.

All of the chapters report a considerable lack of research in parasites and changing parasite distributions and associations within Southeast Asia. In addition, they show that older identification techniques based on morphology are frequently inadequate to identify the parasites of concern. Thus, more recent molecular techniques should be more commonly used, particularly in epidemiological studies, to provide accurate information for prevention and control strategies. Given the significance of parasites for human and animal health in the area, these research deficits should be strongly addressed.

Karlsruhe, Germany — Trevor N. Petney
Khon Kaen, Thailand
Maha Sarakham, Thailand — Weerachai Saijuntha
Düsseldorf, Germany — Heinz Mehlhorn

# Contents

# Contributors

**Ross H. Andrews** Cholangiocarcinoma Research Institute, Faculty of Medicine, Khon Kaen University, Khon Kaen, Thailand
Faculty of Medicine, Imperial College, London, UK

**Kittipong Chaisiri** Department of Helminthology, Faculty of Tropical Medicine, Mahidol University, Bangkok, Thailand

**Pewpan M. Intapan** Department of Parasitology, Faculty of Medicine, Khon Kaen University, Khon Kaen, Thailand
Mekong Health Science Research Institute, Khon Kaen University, Khon Kaen, Thailand

**Wanchai Maleewong** Department of Parasitology, Faculty of Medicine, Khon Kaen University, Khon Kaen, Thailand
Mekong Health Science Research Institute, Khon Kaen University, Khon Kaen, Thailand

**Heinz Mehlhorn** Department of Parasitology, Heinrich Heine University, Düsseldorf, Germany

**Serge Morand** Department of Helminthology, Faculty of Tropical Medicine, Mahidol University, Bangkok, Thailand
CNRS-CIRAD, Faculty of Veterinary Technology, Kasetsart University, Bangkok, Thailand

**Trevor N. Petney** Departments of Zoology and Paleontology and Evolution, State Museum of Natural History, Karlsruhe, Germany
Cholangiocarcinoma Research Institute, Khon Kaen University, Khon Kaen, Thailand

**Pairot Pramual** Department of Biology, Faculty of Science, Mahasarakham University, Maha Sarakham, Thailand

**Richard G. Robbins** Emeritus Civilian Medical Entomologist at the Walter Reed Biosystematics Unit, Department of Entomology, Smithsonian Institution, MSC, MRC 534, Suitland, MD, USA

**Weerachai Saijuntha** Walai Rukhavej Botanical Research Institute, Biodiversity and Conservation Research Unit, Mahasarakham University, Maha Sarakham, Thailand

**Oranuch Sanpool** Department of Parasitology, Faculty of Medicine, Khon Kaen University, Khon Kaen, Thailand
Mekong Health Science Research Institute, Khon Kaen University, Khon Kaen, Thailand

**Paiboon Sithithaworn** Cholangiocarcinoma Research Institute, Faculty of Medicine, Khon Kaen University, Khon Kaen, Thailand

**Tongjit Thanchomnang** Faculty of Medicine, Mahasarakham University, Maha Sarakham, Thailand

**Kotchaphon Vaisusuk** Department of Veterinary Technology and Veterinary Nursing, Faculty of Agricultural Technology, Rajabhat Maha Sarakham University, Maha Sarakham, Thailand

**Hiroshi Yamasaki** Department of Parasitology, National Institute of Infectious Diseases, Ministry of Health, Labour and Welfare, Tokyo, Japan

# Chapter 1
# The Changing Biodiversity of Parasite Hosts in Southeast Asia

**Weerachai Saijuntha and Trevor N. Petney**

**Abstract** Parasites and their hosts form an integrated system with the parasites life cycle being based on maintaining transmission. Any changes within this system are most likely to influence the transmission potential. Species specific parasites in endangered animals will die out with these hosts. Parasites whose host population benefits from environmental changes, such as generalist and synanthropic species, will have transmission advantages due to the increased frequency with which they encounter suitable hosts. Increasing urbanization suggests that synanthropic hosts for parasites affecting humans will increase in prevalence unless suitable changes in pest control and hygiene are implemented, especially in tropical countries. In contrast, parasites of wild animals will become less frequent as their environment is disturbed or destroyed.

**Keywords** Parasites · Hosts · Ecology · Distribution · Population ecology

## 1.1 Introduction

All parasite species require at least one host while most species require one or more intermediate hosts or a vector to complete their life cycle. Any change in the density or dynamics of the host population will invariably lead to changes in the population dynamics of the parasite species involved. In many cases, the control of parasite

W. Saijuntha (✉)
Walai Rukhavej Botanical Research Institute, Biodiversity and Conservation Research Unit, Mahasarakham University, Maha Sarakham, Thailand
e-mail: weerachai.s@msu.ac.th

T. N. Petney
Departments of Zoology and Paleontology and Evolution, State Museum of Natural History, Karlsruhe, Germany

Cholangiocarcinoma Research Institute, Khon Kaen University, Khon Kaen, Thailand



T. N. Petney et al. (eds.), *Biodiversity of Southeast Asian Parasites and Vectors causing Human Disease*, Parasitology Research Monographs 14,
https://doi.org/10.1007/978-3-030-71161-0_1

populations occurs via the deliberate reduction of the host or vector population (Arneberg et al. 1998; Merino et al. 2011). The development of resistance of intermediate hosts or vectors to control mechanisms can lead to a resurgence of the parasitic disease involved (Ranson et al. 2011; Ranson and Lissenden 2016). In order to determine possible future scenarios for the public health relevance of parasitic diseases it is therefore necessary to understand the dynamics of the host population and its interaction with other species within the ecological community involved.

## 1.2 Overall Situation: Changing Patterns of Diversity

Environmentally, Southeast Asia is undergoing massive changes with continually growing human populations (Gaughan et al. 2013; United Nations, Department of Economic and Social Affairs, Population Division 2019), reduced and frequently fragmented natural habitats (Fox et al. 2012; Richards and Friess 2016), increased agricultural areas, in particular monocultures such as rubber and oil palm (Ziegler et al. 2009; Dallinger 2011; Fox et al. 2012), high levels of pollution, for example, via pesticides and fertilizers (Novotny et al. 2010; Chau et al. 2015), as well as global factors such as climate change (see Petney et al. 2021). In addition, vertebrate and invertebrate populations are at risk via hunting for the pet trade, traditional medicines, and bush meat (Harrison et al. 2016). These factors are likely to interact causing synergistic effects that may be difficult to predict but will increase the risk of a reduction in distributional areas and potentially species extinction (Brook et al. 2008). They will lead to overall reductions in biodiversity (Gray et al. 2018).

On the other hand, although the increase in human dominated lands leads to disadvantages for wild living species, there are advantages for synanthropic species due to the reduction in natural diversity due to a reduction in the number of habitats/microhabitats available (Bowen et al. 2007; Alkemade et al. 2009; Guetté et al. 2017). Synanthropic species include murid and other rodents (see Chaisiri and Morand this volume; Morand et al. 2019), bird species, such as the domestic pigeon (Yap and Sodhi 2004), as well as invertebrates such as various fly species that are capable of transporting parasite species (Graczyk et al. 1999, 2005) and mosquito vectors that live in human effluent (Chaves et al. 2009). These are likely to benefit substantially by human presence with the invasion of new areas and increased population size potentially leading to increase in transmission rates.

## 1.3 Terrestrial Vertebrates

The biodiversity of terrestrial vertebrates is declining worldwide at a dramatic rate (WWF 2018, 2020). This is also true for Southeast Asia. The overall estimate for biodiversity loss lies by 13–85% (Sodhi et al. 2010), while for Southeast Asia Sodhi et al. (2004) list potential extinctions of mammals of approximately between 25–60%, birds 15–30%, reptiles 5–25%, and amphibians between 5 and 45% by 2100. This does not consider the high potential for numerous taxa to actually contain groups of cryptic species, which would be likely decrease the estimated distributional area and population size of the individual species within the group (Bickford et al. 2007; Francis et al. 2010; Saijuntha et al. 2017, 2019a, b). Hughes (2017) shows that most of the biodiversity hotspots in Southeast Asia are not incorporated into protected areas.

In contrast, synanthropic species living in close contact with humans (Fig. 1.1) are likely to benefit from human presence. For example, Wenz-Muecke et al. (2013) compared groups of long-tailed macaque (Fig. 1.1a) with frequent contact with human modified environments, including human food. They showed that the intensities of infection with *Strongyloides fuelleborni* and of an intestinal fluke (probably *Haplorchis* sp.) were substantially lower in the sylvatic groups. The authors concluded that the macaque populations in peri-urban habitats change their behavior in such a way as to increase the likelihood of infection with human parasites, potentially increasing the threat to humans by a new zoonotic source. The variable squirrel (Fig. 1.1c) is common in peri-urban and urban areas where it utilizes human constructions such as power lines and building for daily movement (Kobayashi

**Fig. 1.1** Free-ranging animals usually feeding close to human communities; long-tailed macaque (**a**, *Macaca fascicularis*) photo by Weerachai Saijuntha, hog badger (**b**, *Arctonyx collaris*) photo courtesy by Komgrit Wongpakam, variable squirrel (**c**, *Callosciurus finlaysonii*) photo courtesy by Supaporn Teamwong, stray dog (**d**, *Canis familiaris*) photo by Weerachai Saijuntha, wild pig (**e**, *Sus scrofa*) photo courtesy by Ubon Tangkawanit, banteng (**f**, *Bos javanicus*) photo courtesy by Watee Kongbuntad

et al. 2018). It has been introduced to countries external to Southeast Asia, presumably via the pet trade (Oshida et al. 2007; Bertolino et al. 2004; Bertolino and Lurz 2013), where it is known to harbor endoparasites (d'Ovidio et al. 2014). Stray and companion dogs (Fig. 1.1d) are an everyday occurrence in most if not all areas where humans are found. They (as well as domestic cats) also harbor parasites of significance to public health (Hinz 1980; Jittapalapong et al. 2007, 2009). Wild boar (Fig. 1.1e) are common throughout Southeast Asia where they may be involved in peri-urban cycles of parasites such as *Trichinella spiralis* (Pozio 2001; Thi et al. 2014).

In contrast, the hog badger (Fig. 1.1b) is listed as vulnerable on the International Union for the Conservation of Nature Red List (Chutipong et al. 2014; Gray et al. 2018). The banteng (Fig. 1.1f) is listed as endangered (Nguyen 2009; Rahman et al. 2019).

Numerous species of birds are likely to become endangered due to environmental change, particularly habitat destruction (Castelletta et al. 2000; Sodhi 2002) or the pet trade (Sodhi et al. 2006; Harris et al. 2017). Figure 1.2 shows eight species with a very wide distribution. Species A–D are more reclusive and do not frequent human habitats while species E–H are found more commonly on agricultural land on ponds (Robson 2005).

Various species of bird, particularly those associated with freshwater environments, can act as final hosts of parasites relevant to public health (Chai et al. 2009), for example, the intestinal flukes *Haplorchis pumilio* and *H. taichui* both of which have high prevalences in humans in some areas of mainland and island Southeast Asia (De et al. 2003; Belizario Jr et al. 2004; Chai et al. 2007, 2010; Watthanakulpanich et al. 2010).

## 1.4 Reptiles

Worldwide, reptiles are subjected to numerous changes that threaten their existence. Gibbons et al. (2000) list as the most significant threats: habitat loss and degradation, introduced invasive species, environmental pollution, disease, and unsustainable use and climate change.

Southeast Asian reptiles are particularly subject to habitat change, as well as the pet and traditional medicine trades (Koch et al. 2013; Natusch et al. 2019; Marshall et al. 2020). Climate change is likely to cause significant changes to the likelihood of reptile survival. Bickford et al. (2010) hypothesize, based on known physiological and ecological tolerances, that within 50 years reptiles in Southeast Asia will no longer be able to adapt to due to temperature dependent sex determination, higher metabolic rates, and less bio-available water.

A significant problem here is the fact that some common species, also sold as pets and medicines, occur as species complexes that have not yet been fully elucidated. The tokay gecko (*Gekko gecko*) is used as a traditional medicine, particularly in China and Malaysia (Caillabet 2013), for a variety of diseases including AIDS,

**Fig. 1.2** Diversity of some common birds found in Khon Kaen Province, Thailand; brown shrike catching a frog (**a**; *Lanius cristatus*), Brahminy kite catching a fish (**b**; *Haliastur indus*), yellow bittern eating a fish (**c**, *Ixobrychus sinensis*), common kingfisher catching a fish (**d**, *Alcedo atthis*), little grebe catching a fish (**e**, *Tachybaptus ruficollis*), little egret catching a fish (**f**, *Egretta garzetta*), Asian openbill stork eating an apple snail (**g**, *Anastomus oscitans*), grey-headed swamphen catching an apple snail (**h**, *Porphyrio indicus*), photos courtesy by Supaporn Teamwong

cancer, asthma, tuberculosis, diabetes, skin disease, and impotence (Bauer 2009). It is estimated that between 2–5 million dried specimens are exported annually from Thailand (Laoong and Sribundit 2006) and 1.2 million specimens from Java (Nijman et al. 2012). Other specimens come from Lao PDR and Cambodia (Kongbuntad et al. 2016). These tokays are mostly exported to China and Malaysia for use as a traditional Chinese medicine. Work by Saijuntha and colleagues on the tokay gecko (*Gekko gecko*), a species also found in residential areas (Thirakhupt et al. 2006), has demonstrated highly significant genetic differences between populations from different localities and indicates that this taxon comprises a number of species, the detailed distributions and ecologies of which are either only rudimentarily known or not known at all (Kongbuntad et al. 2016; Saijuntha et al. 2019a, b).

## 1.5 Fish

Fish act as the intermediate hosts for a wide variety of parasites in Southeast Asia. Of particular importance are the liver flukes *Opisthorchis viverrini* in the Mekong basin and *Clonorchis sinensis* from central Vietnam northwards, as well as certain species of intestinal fluke (Chai et al. 2009; Saijuntha et al. 2019a, b). As with other host groups there are major dynamic changes taking place in Southeast Asian fish populations, however, these are related to the freshwater rather than the land environment. Flowing freshwater ecosystems in Southeast Asia are being exploited by the construction of many dams for water catchment both as reliable sources of water for human use as well as for generating hydroelectric power (Grumbine and Xu 2011; Hecht et al. 2019). This has a major impact on the environment including human livelihoods (Sovacool and Bulan 2011). Fish represent the major source of protein for much of the Southeast Asian human population with a reduction in the fish catch being predicted after dam construction (Orr et al. 2012).

Many species of fish in the Mekong River are migratory (Poulsen et al. 2002) with dams severely restricting or completely cutting off this annual movement (Dugan et al. 2010; Ziv et al. 2012; Stone 2016; Golden et al. 2019). In addition, water level, which is strongly correlated with discharge and water current, appears to be the major trigger for migration. The thresholds or change in water level, discharge or current trigger the migration of 30 species of Mekong fish (Baran 2006).

*Opisthorchis viverrini* is the major risk factor for developing biliary cancer (cholangiocarcinoma) in those countries bordering the Mekong River in Southeast Asia. This is a disease of major public health significance with very high mortality rates, particularly in the rural populations of northern and northeastern Thailand and Lao PDR (Khuntikeo et al. 2018). Transmission occurs through the ingestion of raw or undercooked cyprinid fish (Fig. 1.3) containing the parasite's infective metacercariae (Grundy-Warr et al. 2012).

## 1.6 Invertebrates

The conservation status of invertebrates is less well understood than that of vertebrate species, although invertebrates, especially those with a freshwater component to their life cycle, frequently act as intermediate hosts or vectors for parasites of public health significance. Indications are that environmental change will lead to major changes in freshwater ecosystems in tropical Southeast Asia (Dudgeon 1992, 2000; Giam et al. 2010). In addition, invertebrate species, such as house flies, can act as transfer hosts for parasite species without being necessary for the parasite's life cycle. Zieritz et al. (2018) found that seven unionid mussel subfamilies are native East and Southeast Asia with species richness being highest in Southeast Asia for four species. Only 24% (61) of species have a known conservation status post-1980, with data for the Philippines, Laos, Indonesia, Myanmar, and Malaysia being

**Fig. 1.3** Diversity of cyprinid fish second intermediate hosts of *Opisthorchis viverrini* found in Chi River in Mekong Basin, Maha Sarakham Province, Thailand; Smith's barb or Pla Mang (in Thai) (**a**, *Puntioplites proctozystron*), Siamese mud carp or pla soi khao (**b**, *Henicorhynchus siamensis*), red tailed tinfoil (**c**, *Barbodes altus*), Java barb or ta-phian (**d**, *Barbodes gonionotus*), pla sood (**e**, *Hampala dispar*), tinfoil barb (**f**, *Barbodes schwanenfeldii*), photos courtesy by Komgrit Wongpakam

particularly limited. The authors indicate that habitat modification, including the damming of rivers, and increasing pollution levels are likely to act as major threats to the maintenance of populations.

Notable vectors in Southeast Asia are *Anopheles* species transmitting malaria (Trung et al. 2004), and as the main vector of filariasis *Culex quinquefasciatus*, with a more limited contribution by *Aedes* and *Mansonia* spp. (Dickson et al. 2017). These are major diseases causing significant morbidity and mortality throughout the region (WHO 2016; Barber et al. 2017; Dickson et al. 2017). Environmental change can act in two ways: *Anopheles dirus*, one on the main vectors of malaria on mainland Southeast Asia, is a sylvatic species requiring shady, slow flowing streams for egg laying. Forest destruction in Thailand has substantially reduced the distributional area of malaria, especially in the northeast of the country, as suitable habitats for this vector are missing (Petney et al. 2009). On the other hand, certain species such as *Culex quinquefasciatus* is decidedly synanthropic breeding in water, including effluent, with a high organic content (Easton 1994).

Parasite transfer occurs when the species carrying the parasite to a new host is not a necessary component in the parasites life cycle. Thus various species of Diptera can transfer human pathogens (Sallehudin et al. 2000; Baldacchino et al. 2013; Khamesipour et al. 2018). For the domestic fly, *Musca domestica*, Khamesipour and colleagues list 1 species of amoeba and 11 genera of helminths containing at least 15 species that are transferred by this species. The transfer of these parasites within urban and peri-urban areas is likely to be facilitated by this species.

## 1.7 Animal Trade

In addition to changes in the environment, there is a massive trade in wild vertebrates and invertebrates (Nijman 2010; OECD 2019; Scheffers et al. 2019; Krishnasamy and Zavagli 2020) for use in traditional medicine (Lee et al. 2014; Fig. 1.4a), the pet trade (Bush et al. 2014) and, in some areas, bush meat (Lee et al. 2014; Fig. 1.4b, c). Figure 1.4d illustrates birds captured for release at Buddhist temples as a means of gaining merit. Harrison et al. (2016) list as the proximate causes for the increased hunting of wild animals, improved access to forests and markets, improved technology, and increases in the demand for bush meat, products for tradition medicine, and the pet trade. They conclude that hunting poses the most significant, direct threat to endangered vertebrate survival in Southeast Asia, and that the current trend is non-sustainable.

## 1.8 Effect on Parasites

Parasites show a wide spectrum of host usage from species infecting only a single or few hosts, for example, the tick *Amblyomma crenatum* occurs only on rhinoceros, to more generalist species such as *Rhipicephalus microplus* occurring predominantly on members of the family Bovidae, to various species of the tick genus *Ixodes* (e.g., *Ixodes ricinus*) that can parasitize a wide range on mammal, bird, and reptile hosts (Guglielmone et al. 2014). Here, host specificity can be severely detrimental to wild, endangered hosts (Saijuntha et al. this volume) but advantageous to synanthropic species. Thus *Rhipicephalus sanguineus sensu lato*, a parasite predominantly of domestic dogs, can be found in association with humans and their canine pets worldwide (Guglielmone et al. 2014). A reduction in the wild host spectrum or population size will reduce the population size and likely survival of a parasite species as the probability of encountering a host, and in the case of adults, a host with a potential mate will be reduced. Once a lower threshold is reached not enough hosts will be available for parasite survival. The more host specific the parasite, the less likely it is to survive.

**Fig. 1.4** Animal trade in Thailand: tokay geckos dried for export as Chinese traditional medicine (**a**), moles for sale for cooking in a local market (**b**), turtles (**c**) and scaly-breasted munia (**d**, *Lonchura punctulata*) for sale to be released for making merit in a temple, (photos courtesy by Warayutt Pilap)

In the case of synanthropic hosts, it is likely that the host population size will be at least stable (if no control measures are undertaken) or increase depending on the extent of human influence on the environment. This will increase the likelihood of parasite transmission.

## 1.9 Consequences for human and Animal Health

Humans and their domestic and stock animals can be affected by: (1) parasites that prefer synanthropic hosts are more likely to be transmitted due to their association with humans (Pozio 2000; Chaisiri and Morand this volume); and (2) human encroachment on wild habitats is likely to bring people into contact with novel and potentially dangerous parasites to which they have no acquired immunity (Patz et al. 2000; Thompson 2013; Mackenstedt et al. 2015); (3) parasite switching or concentration in hosts likely to contact humans by the elimination of other wild hosts (Pfäffle et al. 2015).

The most likely hosts of parasites that can potentially affect humans and their companion and stock animals are other mammals. It is possible the closer the relationship the more likely that transmission will occur. For example, the malaria parasites *Plasmodium knowlesi* and *P. cynomolgi* infect primate species from Southeast Asia. Both are now known to switch to human hosts presumably via close contact with their natural hosts (Vythilingam et al. 2008; Imwong et al. 2019).

## References


Alkemade R, Van Oorschot M, Miles L, Nellemann C, Bakkenes M, Ten Brink B (2009) GLOBIO3: a framework to investigate options for reducing global terrestrial biodiversity loss. Ecosystems 12(3):374–390

Arneberg P, Skorping A, Grenfell B, Read AF (1998) Host densities as determinants of abundance in parasite communities. Proc R Soc Lond Ser B Biol Sci 265(1403):1283–1289

Baldacchino F, Muenworn V, Desquesnes M, Desoli F, Charoenviriyaphap T, Duvallet G (2013) Transmission of pathogens by *Stomoxys* flies (Diptera, Muscidae): a review. Parasite 2013:20

Baran E (2006) Fish migration triggers in the Lower Mekong Basin and other tropical freshwater systems. MRC Technical Paper No. 14, Mekong River Commission, Vientiane, p 56

Barber BE, Rajahram GS, Grigg MJ, William T, Anstey NM (2017) World malaria report: time to acknowledge *Plasmodium knowlesi* malaria. Malar J 16(1):1–3

Bauer AM (2009) Geckos in traditional medicine: forensic implications. Appl Herpetol 6(1):81–96

Belizario VY Jr, de Leon WU, Bersabe MMJ, Baird JK, Bangs MJ (2004) A focus of human infection by *Haplorchis taichui* (Trematoda: Heterophyidae) in the Southern Philippines. J Parasitol 90(5):1165–1169

Bertolino S, Lurz PWW (2013) *Callosciurus* squirrels: worldwide introductions, ecological impacts and recommendations to prevent the establishment of new invasive populations. Mammal Rev 43:22–33

Bertolino S, Mazzoglio PJ, Vaiana M, Currado I (2004) Activity budget and foraging behavior of introduced *Callosciurus finlaysonii* (Rodentia, Sciuridae) in Italy. J Mammal 85(2):254–259

Bickford D, Lohman DJ, Sodhi NS, Ng PK, Meier R, Winker K, Ingram KK, Das I (2007) Cryptic species as a window on diversity and conservation. Trends Ecol Evol 22(3):148–155

Bickford D, Howard SD, Ng DJ, Sheridan JA (2010) Impacts of climate change on the amphibians and reptiles of Southeast Asia. Biodivers Conserv 19(4):1043–1062

Bowen ME, McAlpine CA, House AP, Smith GC (2007) Regrowth forests on abandoned agricultural land: a review of their habitat values for recovering forest fauna. Biol Conserv 140 (3-4):273–296

Brook BW, Sodhi NS, Bradshaw CJ (2008) Synergies among extinction drivers under global change. Trends Ecol Evol 23(8):453–460

Bush ER, Baker SE, Macdonald DW (2014) Global trade in exotic pets 2006–2012. Conserv Biol 28(3):663–676

Caillabet OS (2013) The trade in tokay geckos *Gekko gecko* in South-east Asia: with a case study on novel medical claims in peninsular Malaysia. TRAFFIC, Petaling Jaya

Castelletta M, Sodhi NS, Subaraj R (2000) Heavy extinctions of forest avifauna in Singapore: lessons for biodiversity conservation in Southeast Asia. Conserv Biol 14(6):1870–1880

Chai JY, Han ET, Guk SM, Shin EH, Sohn WM, Yong TS, Eom KS, Lee KH, Jeong HG, Ryang YS, Hoang EH (2007) High prevalence of liver and intestinal fluke infections among residents of Savannakhet Province in Laos. Korean J Parasitol 45(3):213

Chai JY, Shin EH, Lee SH, Rim HJ (2009) Foodborne intestinal flukes in Southeast Asia. Korean J Parasitol 47(Suppl):S69

Chai JY, Yong TS, Eom KS, Min DY, Shin EH, Banouvong V, Insisiengmay B, Insisiengmay S, Phommasack B, Rim HJ (2010) Prevalence of the intestinal flukes *Haplorchis taichui* and *H. yokogawai* in a mountainous area of Phongsaly Province, Lao PDR. Korean J Parasitol 48 (4):339

Chau ND, Sebesvari Z, Amelung W, Renaud FG (2015) Pesticide pollution of multiple drinking water sources in the Mekong Delta, Vietnam: evidence from two provinces. Environ Sci Pollut Res 22(12):9042–9058

Chaves LF, Keogh CL, Vazquez-Prokopec GM, Kitron UD (2009) Combined sewage overflow enhances oviposition of *Culex quinquefasciatus* (Diptera: Culicidae) in urban areas. J Med Entomol 46(2):220–206

Chutipong W, Tantipisanuh N, Ngoprasert D, Lynam AJ, Steinmetz R, Jenks KE, Sukmasuang GG, Harichu FK, Tredyte AC, Cutter P, Cutter PB (2014) Current distribution and conservation status of small carnivores in Thailand: a baseline review. Small Carniv Conserv 51:96–136

d'Ovidio D, Rinaldi L, Ianniello D, Donnelly TM, Pepe P, Capasso M, Cringoli G (2014) FLOTAC for diagnosis of endo-parasites in pet squirrels in southern Italy. Vet Parasitol 200(1-2):221–214

Dallinger J (2011) Oil palm development in Thailand: economic, social and environmental considerations. Oil Palm Expans South East Asia 2011:24–51

De NV, Murrel KD, Cong le D (2003) The food-borne trematode zoonoses of Vietnam. Southeast Asian J Trop Med Public Health 34(suppl 1):12–34

Dickson BF, Graves PM, McBride WJ (2017) Lymphatic filariasis in mainland Southeast Asia: a systematic review and meta-analysis of prevalence and disease burden. Trop Med Inf Dis 2 (3):32

Dudgeon D (1992) Endangered ecosystems: a review of the conservation status of tropical Asian rivers. Hydrobiologia 248(3):167–191

Dudgeon D (2000) The ecology of tropical Asian rivers and streams in relation to biodiversity conservation. Annu Rev Ecol Syst 31(1):239–263

Dugan PJ, Barlow C, Agostinho AA, Baran E, Cada GF, Chen D, Cowx IG, Ferguson JW, Jutagate T, Mallen-Cooper M, Marmulla G (2010) Fish migration, dams, and loss of ecosystem services in the Mekong basin. Ambio 39(4):344–348

Easton ER (1994) Urbanization and its effects on the ecology of mosquitoes in Macau, Southeast Asia. J Am Mosq Control Assoc 10(4):540–544

Fox J, Vogler JB, Sen OL, Giambelluca TW, Ziegler AD (2012) Simulating land-cover change in montane mainland southeast Asia. Environ Manag 49(5):968–979

Francis CM, Borisenko AV, Ivanova NV, Eger JL, Lim BK et al (2010) The role of DNA barcodes in understanding and conservation of mammal diversity in Southeast Asia. PLoS One 5(9): e12575

Gaughan AE, Stevens FR, Linard C, Jia P, Tatem AJ (2013) High resolution population distribution maps for Southeast Asia in 2010 and 2015. PLoS One 8(2):e55882

Giam X, Ng TH, Yap VB, Tan HT (2010) The extent of undiscovered species in Southeast Asia. Biodivers Conserv 19(4):943–954

Gibbons JW, Scott DE, Ryan TJ, Buhlmann KA, Tuberville TD, Metts BS, Greene JL, Mills T, Leiden Y, Poppy S, Winne CT (2000) The global decline of reptiles, Déjà vu amphibians: reptile species are declining on a global scale. Six significant threats to reptile populations are habitat loss and degradation, introduced invasive species, environmental pollution, disease, unsustainable use, and global climate change. Bioscience 50(8):653–666

Golden CD, Shapero A, Vaitla B, Smith MR, Myers SS, Stebbins EA, Gephart JA (2019) Impacts of hydropower development on fisheries and human nutrition in the Lower Mekong. Front Sust Food Syst 3:93

Graczyk TK, Cranfield MR, Fayer R, Bixler H (1999) House flies (*Musca domestica*) as transport hosts of *Cryptosporidium parvum*. Am J Trop Med Hyg 61(3):500–504

Graczyk TK, Knight R, Tamang L (2005) Mechanical transmission of human protozoan parasites by insects. Clin Microbiol Rev 18(1):128–132

Gray TN, Hughes AC, Laurance WF, Long B, Lynam AJ, O'Kelly H, Ripple WJ, Seng T, Scotson L, Wilkinson NM (2018) The wildlife snaring crisis: an insidious and pervasive threat to biodiversity in Southeast Asia. Biodivers Conserv 27(4):1031–1037

Grumbine RE, Xu J (2011) Mekong hydropower development. Science 332(6026):178–179

Grundy-Warr C, Andrews RH, Sithithaworn P, Petney TN, Sripa B, Laithavewat L, Ziegler AD (2012) Raw attitudes, wetland cultures, life-cycles: socio-cultural dynamics relating to *Opisthorchis viverrini* in the Mekong Basin. Parasitol Int 61(1):65–70

Guetté A, Gaüzère P, Devictor V, Jiguet F, Godet L (2017) Measuring the synanthropy of species and communities to monitor the effects of urbanization on biodiversity. Ecol Indic 79:139–154

Guglielmone AA, Robbins RG, Apanaskevich DA, Petney TN, Estrada-Peña A, Horak IG (2014) The hard ticks of the world. Springer, New York

Harris JB, Tingley MW, Hua F, Yong DL, Adeney JM, Lee TM, Marthy W, Prawiradilaga DM, Sekercioglu CH, Winarni N, Wilcove DS (2017) Measuring the impact of the pet trade on Indonesian birds. Conserv Biol 31(2):394–405

Harrison RD, Sreekar R, Brodie JF, Brook S, Luskin M, O'Kelly H, Rao M, Scheffers B, Velho N (2016) Impacts of hunting on tropical forests in Southeast Asia. Conserv Biol 30(5):972–981

Hecht JS, Lacombe G, Arias ME, Dang TD, Piman T (2019) Hydropower dams of the Mekong River basin: a review of their hydrological impacts. J Hydrol 568:285–300

Hinz E (1980) Intestinal helminths in Bangkok stray dogs and their role in public health (author's tranls). Zentralbl Bakteriol Mikrobiol Hygiene 1. Abt Orig B Hyg 171(1):79–85

Hughes AC (2017) Mapping priorities for conservation in Southeast Asia. Biol Conserv 209:395–405

Imwong M, Madmanee W, Suwannasin K, Kunasol C, Peto TJ, Tripura R, Seidlein L, Nguon C, Davoeung C, Day NP, Dondorp AM (2019) Asymptomatic natural human infections with the simian malaria parasites *Plasmodium cynomolgi* and *Plasmodium knowlesi*. J Informet 219 (5):695–702

Jittapalapong S, Nimsupan B, Pinyopanuwat N, Chimnoi W, Kabeya H, Maruyama S (2007) Seroprevalence of *Toxoplasma gondii* antibodies in stray cats and dogs in the Bangkok metropolitan area, Thailand. Vet Parasitol 145(1-2):138–141

Jittapalapong S, Sittisan P, Sakpuaram T, Kabeya H, Maruyama S, Inpankaew T (2009) Coinfection of *Leptospira* spp. and *Toxoplasma gondii* among stray dogs in Bangkok, Thailand. Southeast Asian J Trop Med Public Health 40(2):247

Khamesipour F, Lankarani KB, Honarvar B, Kwenti TE (2018) A systematic review of human pathogens carried by the housefly (*Musca domestica* L.). BMC Public Health 18(1):1049

Khuntikeo N, Titapun A, Loilome W, Yongvanit P, Thinkhamrop B, Chamadol N, Boonmars T, Nethanomsak T, Andrews RH, Petney TN, Sithithaworn P (2018) Current perspectives on opisthorchiasis control and cholangiocarcinoma detection in Southeast Asia. Front Med 5:117

Kobayashi S, Placksanoi J, Taksin A, Aryuthaka C, Denda T, Izawa M (2018) Resource use among sympatric *Callosciurus* spp. (Sciuridae) in tropical seasonal forests during the dry season in northeastern Thailand. Mamm Stud 44(1):23–32

Koch A, Ziegler T, Boehme W, Arida E, Auliya M (2013) Pressing problems: distribution, threats, and conservation status of the monitor lizards (Varanidae: *Varanus* spp.) of Southeast Asia and the Indo-Australian Archipelago. Herpetol Conserv Biol 8(3):1–62

Kongbuntad W, Tantrawatpan C, Pilap W, Jongsomchai K, Chanaboon T, Laotongsan P, Petney TN, Saijuntha W (2016) Genetic diversity of the red-spotted tokay gecko (*Gekko gecko* Linnaeus, 1758) (Squamata: Gekkonidae) in Southeast Asia determined with multilocus enzyme electrophoresis. J Asia-Pac Biodivers 9(1):63–68

Krishnasamy K, Zavagli M (2020) Southeast Asia: at the heart of wildlife trade. TRAFFIC, Petaling Jaya

Laoong S, Sribundit W (2006) Diet of Tokay gecko (*Gekko gecko*) in Eastern and Northern regions of Thailand. Wildlife Yearbook 7:78–90

Lee TM, Sigouin A, Pinedo-Vasquez M, Nasi R (2014) The harvest of wildlife for bushmeat and traditional medicine in East, South and Southeast Asia: Current knowledge base, challenges, opportunities and areas for future research. Occasional Paper 115. CIFOR, Bogor

Mackenstedt U, Jenkins D, Romig T (2015) The role of wildlife in the transmission of parasitic zoonoses in peri-urban and urban areas. Int J Parasitol 4(1):71–79

Marshall BM, Strine C, Hughes AC (2020) Thousands of reptile species threatened by under-regulated global trade. Nat Commun 11(1):1–2

Merino O, Almazán C, Canales M, Villar M, Moreno-Cid JA, Galindo RC, De la Fuente J (2011) Targeting the tick protective antigen subolesin reduces vector infestations and pathogen infection by *Anaplasma marginale* and *Babesia bigemina*. Vaccine 29(47):8575–8579

Morand S, Blasdell K, Bordes F, Buchy P, Carcy B, Chaisiri K, Chaval Y, Claude J, Cosson JF, Desquesnes M, Jittapalapong S (2019) Changing landscapes of Southeast Asia and rodent-borne diseases: decreased diversity but increased transmission risks. Ecol Appl 29(4):e01886

Natusch DJ, Lyons JA, Riyanto A, Khadiejah S, Shine R (2019) Detailed biological data are informative, but robust trends are needed for informing sustainability of wildlife harvesting: a case study of reptile offtake in Southeast Asia. Biol Conserv 233:83–92

Nguyen MH (2009) The status of vulnerable gaur *Bos gaurus* and endangered banteng *Bos javanicus* in Ea So Nature Reserve and Yok Don and Cat Tien national parks, Vietnam. Oryx 43(1):129–135

Nijman V (2010) An overview of international wildlife trade from Southeast Asia. Biodivers Conserv 19(4):1101–1114

Nijman V, Todd M, Shepherd CR (2012) Wildlife trade as an impediment to conservation as exemplified by the trade in reptiles in Southeast Asia. In: Gower DJ, Johnson KG, Richardson JE, Rosen BR, Rüber L, Williams ST (eds) Biotic evolution and environmental change in Southeast Asia. Cambridge University Press, Cambridge

Novotny V, Wang X, Englande AJ, Bedoya D, Promakasikorn L, Tirado R (2010) Comparative assessment of pollution by the use of industrial agricultural fertilizers in four rapidly developing Asian countries. Environ Dev Sustain 12(4):491–509

OECD (2019) The illegal wildlife trade in Southeast Asia: institutional capacities in Indonesia, Singapore, Thailand and Viet Nam, illicit trade. OECD Publishing, Paris

Orr S, Pittock J, Chapagain A, Dumaresq D (2012) Dams on the Mekong River: lost fish protein and the implications for land and water resources. Glob Environ Chang 22(4):925–932

Oshida T, Torii H, Lin LK, Lee JK, Chen YJ, Endo H, Sasaki M (2007) A preliminary study on origin of *Callosciurus* squirrels introduced into Japan. Mamm Stud 32(2):75–82

Patz JA, Graczyk TK, Geller N, Vittor AY (2000) Effects of environmental change on emerging parasitic diseases. Int J Parasitol 30(12-13):1395–1405

Petney T, Sithithaworn P, Satrawaha R, Grundy-Warr C, Andrews R, Wang YC, Feng CC (2009) Potential malaria reemergence, northeastern Thailand. Emerg Infect Dis 15(8):1330

Petney TN, Sithithaworn P, Andrews RH (2021) Parasite diversity and dynamics and climate change. In: Petney T, Saijuntha W, Mehlhorn H (eds) Biodiversity of Southeast Asian parasites and vectors causing human disease. Springer International Publishing, Cham

Pfäffle MP, Littwin N, Petney TN (2015) The relationship between biodiversity and disease transmission risk. Res Rep Biodivers Stud 4:9–20

Poulsen AF, Poeu O, Viravong S, Suntornratana U, Tung NT. 2002. Fish migrations of the Lower Mekong River Basin: implications for development, planning and environmental management. MRC Technical Paper No. 8, Mekong River Commission, Phnom Penh, p 62

Pozio E (2000) Factors affecting the flow among domestic, synanthropic and sylvatic cycles of *Trichinella*. Vet Parasitol 93(3-4):241–262

Pozio E (2001) Taxonomy of *Trichinella* and the epidemiology of infection in the Southeast Asia and Australian regions. Southeast Asian J Trop Med Public Health 32:129–132

Rahman DA, Herliansyah R, Rianti P, Rahmat UM, Firdaus AY, Syamsudin M (2019) Ecology and conservation of the endangered banteng (*Bos javanicus*) in Indonesia tropical lowland forest. HAYATI. J Biosci 26(2):68

Ranson H, Lissenden N (2016) Insecticide resistance in African *Anopheles* mosquitoes: a worsening situation that needs urgent action to maintain malaria control. Trends Parasitol 32 (3):187–196

Ranson H, N'guessan R, Lines J, Moiroux N, Nkuni Z, Corbel V (2011) Pyrethroid resistance in African anopheline mosquitoes: what are the implications for malaria control? Trends Parasitol 27(2):91–98

Richards DR, Friess DA (2016) Rates and drivers of mangrove deforestation in Southeast Asia, 2000–2012. Proc Natl Acad Sci 113(2):344–349

Robson C (2005) Birds of Southeast Asia. Princeton University Press, Princeton

Saijuntha W, Khumkratok S, Wongpakam K, Thanonkeo S, Senakhun C, Appamaraka S, Yodsiri S, Thongnetr W, Pilap W, Kongbuntad W, Tawong W, Agatsuma T, Petney TN, Tantrawatpan C (2017) Genetic diversity and population structure of blue-crested lizard, *Calotes mystaceus* Duméril & Bibron, 1837 (Squamata: Agamidae) in Thailand. J Genet 96 (2):377–382

Saijuntha W, Sedlak S, Agatsuma T, Jongsomchai K, Pilap W, Kongbuntad W, Tawong W, Suksavate W, Petney TN, Tantrawatpan C (2019a) Genetic structure of the red-spotted tokay gecko, *Gekko gecko* (Linnaeus, 1758) (Squamata: Gekkonidae) from Mainland Southeast Asia. Asian Herpetol Res 10(2):69–78

Saijuntha W, Sithithaworn P, Kiatsopit N, Andrews RH, Petney TN (2019b) Liver flukes: *Clonorchis* and *Opisthorchis*. In: Toledo R, Fried B (eds) Digenetic trematodes, 2nd edn. Springer, Heidelberg, pp 153–199

Sallehudin S, Mohd ZO, Abdul HA (2000) Isolations of enteric pathogens from synanthropic flies trapped in downtown Kuala Lumpur. J Vector Ecol 25(1):90–93

Scheffers BR, Oliveira BF, Lamb I, Edwards DP (2019) Global wildlife trade across the tree of life. Science 366(6461):71–76

Sodhi NS (2002) A comparison of bird communities of two fragmented and two continuous Southeast Asian rainforests. Biodivers Conserv 11(6):1105–1119

Sodhi NS, Koh LP, Brook BW, Ng PK (2004) Southeast Asian biodiversity: an impending disaster. Trends Ecol Evol 19(12):654–660

Sodhi NS, Pin Koh L, Brook BW (2006) Southeast Asian birds in peril. Auk 123(1):275–277

Sodhi NS, Posa MR, Lee TM, Bickford D, Koh LP, Brook BW (2010) The state and conservation of Southeast Asian biodiversity. Biodivers Conserv 19(2):317–328

Sovacool BK, Bulan LC (2011) Behind an ambitious megaproject in Asia: the history and implications of the Bakun hydroelectric dam in Borneo. Energy Policy 39(9):4842–4859

Stone R (2016) Dam-building threatens Mekong fisheries. Science 354:1084–1085

Thi NV, Nguyen VD, Praet N, Claes L, Gabriël S, Huyen NT, Dorny P (2014) *Trichinella* infection in wild boars and synanthropic rats in northwest Vietnam. Vet Parasitol 200(1-2):207–211

Thirakhupt K, Aowphol A, Voris H, Nabhitabhata J (2006) Foraging ecology of the Tokay gecko, Gekko gecko in a residential area in Thailand. Amphibia-Reptilia 27(4):491–503

Thompson RA (2013) Parasite zoonoses and wildlife: one health, spillover and human activity. Int J Parasitol 43(12-13):1079–1088

Trung HD, Van Bortel W, Sochantha T, Keokenchanh K, Quang NT, Cong LD, Coosemans M (2004) Malaria transmission and major malaria vectors in different geographical areas of Southeast Asia. Tropical Med Int Health 9(2):230–207

United Nations, Department of Economic and Social Affairs, Population Division (2019) World population prospects 2019: data Booklet (ST/ESA/SER.A/424)

Vythilingam I, NoorAzian YM, Huat TC, Jiram AI, Yusri YM, Azahari AH, NorParina I, NoorRain A, LokmanHakim S (2008) *Plasmodium knowlesi* in humans, macaques and mosquitoes in peninsular Malaysia. Parasit Vectors 1(1):26

Watthanakulpanich D, Waikagul J, Maipanich W, Nuamtanong S, Sanguankiat S, Pubampen S, Praevanit R, Mongkhonmu S, Nawa Y (2010) *Haplorchis taichui* as a possible etiologic agent of irritable bowel syndrome-like symptoms. Korean J Parasitol 48(3):225

Wenz-Muecke A, Sithithaworn P, Petney TN, Taraschewski H (2013) Human contact influences the foraging behaviour and parasite community in long-tailed macaques. Parasitology 140 (6):709

WHO (2016) World malaria report 2016. World Health Organization, Geneva

WWF (2018) Living planet report - 2018: aiming higher. WWF, Gland

WWF (2020) Living Planet Report 2020-Bending the curve of biodiversity loss. WWF, Gland

Yap CA, Sodhi NS (2004) Southeast Asian invasive birds: ecology, impact and management. Ornithol Sci 3(1):57–67

Ziegler AD, Fox JM, Xu J (2009) The rubber juggernaut. Science 324(5930):1024–1025

Zieritz A, Bogan AE, Froufe E, Klishko O, Kondo T, Kovitvadhi U, Kovitvadhi S, Lee JH, Lopes-Lima M, Pfeiffer JM, Sousa R (2018) Diversity, biogeography and conservation of freshwater mussels (Bivalvia: Unionida) in East and Southeast Asia. Hydrobiologia 810(1):29–44

Ziv G, Baran E, Nam S, Rodríguez-Iturbe I, Levin SA (2012) Trading-off fish biodiversity, food security, and hydropower in the Mekong River Basin. Proc Natl Acad Sci 109(15):5609–5614

# Chapter 2
# Species Richness and Species Co-occurrence of Helminth Parasites in the *Rattus rattus*-Complex Across Stratified Habitat Landuse Types in Mainland Southeast Asia

**Kittipong Chaisiri and Serge Morand**

**Abstract** This chapter explores some aspects of parasite ecology of helminth infection in the *Rattus rattus*-complex from mainland Southeast Asian countries. Diversity and species association patterns of helminths were assessed in relation to human landuse habitats. A substantial helminth diversity remains undiscovered, with at least 32 parasite species being found. We also discuss potential zoonotic species. Human landuse alteration, i.e. peridomestic habitat, appears to be a hotspot for helminth diversity. This trend is also associated with behavioral traits of the host species which commonly represent habitat generalists. We also examine the presence of helminth versus helminth association patterns using a species co-occurrence probabilistic model. Similar to previous findings, positive associations were more prominent than negative, although persistent negative associations occurred between some certain nematode taxa. Understanding the associated factors influencing parasite colonization, diversity and species association patterns is important in parasite ecology research. This knowledge could provide valuable information for the development of predictive models in disease ecology based on host traits and host–parasite or parasite–parasite interactions, as well as geographical and ecological parameters, particularly landuse alteration by human activities in the Anthropocene.

**Keywords** Helminths · Rodents · Species co-occurrence · Habitat · Southeast Asia

---

K. Chaisiri
Department of Helminthology, Faculty of Tropical Medicine, Mahidol University, Bangkok, Thailand
e-mail: kittipong.cha@mahidol.ac.th

S. Morand (✉)
Department of Helminthology, Faculty of Tropical Medicine, Mahidol University, Bangkok, Thailand

CNRS-CIRAD, Faculty of Veterinary Technology, Kasetsart University, Bangkok, Thailand
e-mail: serge.morand@univ-montp2.fr; serge.morand@cirad.fr



T. N. Petney et al. (eds.), *Biodiversity of Southeast Asian Parasites and Vectors causing Human Disease*, Parasitology Research Monographs 14,
https://doi.org/10.1007/978-3-030-71161-0_2

## 2.1 Studies on Murid Rodents and their Helminth Fauna from Mainland Southeast Asia

Due to their excellent adaptability to different environments, rodents are one of the most diverse groups of mammals on the planet. They represent over 450 genera and more than 2200 species, covering nearly one third of the current mammalian diversity (Wilson and Reeder 2005; Burgin et al. 2018). Among the Rodentia, murid rodents (Family Muridae: Old World rats and mice) are the most diversified taxon (Fabre et al. 2012). Rodents occupy nearly all terrestrial ecosystems in all continents (with the exception of Antarctica and excluding invasive rats and mice), from fertile areas (e.g., primary and secondary forests), human land alteration (e.g., agricultural fields and cities), to harsh landscapes (e.g., deserts and frozen lands). They are usually thought of as threats to humans or pests, and are also recognized as prime disease carriers of several zoonotic parasites and pathogens (Morand et al. 2006; Meerburg et al. 2009; Centers for Disease Control and Prevention 2017). Meerburg et al. (2009) presented an extensive review of over 60 pathogens of public health importance that are carried by rodents, i.e. several species of microparasites (viruses, bacteria, and protists) and macroparasites (helminths).

In Southeast Asia, murid rodents are also present in a wide range of habitats including forests, grassland, agricultural areas, and human settlements, and some species are commonly found in large cities, on streets, in fresh food markets, and public parks (Aplin et al. 2003; Blasdell et al. 2015; Stuart et al. 2015; Paladsing et al. 2020). Palmeirim et al. (2014) and Blasdell et al. (2015) reported several patterns of habitat preferences for murid rodents in Southeast Asia using published data from studies in Thailand, Lao PDR, and Cambodia. Some rodent species show specific habitat preferences, e.g. *Rattus norvegicus* and *R. exulans* prefer human build-up habitats (synanthropic species); *R. argentiventer*, *R. sakaeratensis*, and *Bandicota indica* prefer lowland rice fields; and *Maxomys surifer* and *Leopoldamys edwardsi* in forest areas. On the other hand, some species show a more generalist ecology with no clear habitat preference. They tend to be present in more than one habitat type, e.g. *Niviventer fulvescens*, *Berylmys berdmorei*, and *B. bowersi* occur between forest and upland agricultural fields. There is one exceptional rodent species, *R. tanezumi* (black rat or Oriental house rat) that has been reported as a habitat generalist (low habitat preference), occurring in all types of habitats. Regarding the issue of cryptic species among the black rat complex, mitochondrial DNA markers separate this problematic taxon into the three lineages: *R. rattus* (R1), *R. tanezumi* (R2), and unnamed *Rattus* species phylogenetic R3 (Pagès et al. 2010). On the Indochinese Peninsula, only *R. tanezumi* and *Rattus* species phylogenetic R3 have been recorded so far. The two lineages are recognized as the Asian black rats, whereas the true *R. rattus* (R1) has never been found in the region based on genetic markers (Pagès et al. 2010; Aplin et al. 2011; Paladsing et al. 2020). Here, we will apply the term "*Rattus rattus*- complex" as representative for these lineages throughout this study. Because the species has adapted well to the changing environment of Southeast Asia (Chaisiri et al. 2015; Morand et al. 2015), *R. tanezumi* may potentially act as a "bridge species" hosting parasites and pathogens that can spread from one habitat to others.

Helminthiases are parasitic diseases caused by macroparasitic worms (i.e., cestodes, trematodes, nematodes, and acanthocephalans) that are frequently considered as neglected diseases. To date, there are a growing number of studies on the ecology of parasitic infections in small mammals, particularly the host species living close to humans, such as rodents. A number of potential zoonotic helminth species carried by rodents have been documented, e.g. trematodes: *Echinostoma* spp., *Plagiorchis muris*; cestodes: *Hymenolepis diminuta*, *Hymenolepis nana*, *Raillietina* spp.; nematodes *Angiostrongylus cantonensis*, *Capillaria hepatica* (syn. *Calodium hepaticum*), *Cyclodontostomum purvisi*; and the acanthocephalan: *Moniliformis moniliformis* (Chaisiri et al. 2015; McGarry et al. 2015; Jarvi et al. 2017; Ranjbar et al. 2017). Apart from these zoonotic species, a great diversity of rodent specific helminths has also been widely reported. Almost 60 helminth species have been documented in Southeast Asia, with 13 species of cestodes, 15 species of trematodes, about 30 species of nematodes, and one species of acanthocephalan (Wiroreno 1978; Pham et al. 2001; Claveria et al. 2005; Paramasvaran et al. 2005; Chaisiri et al. 2012; Mohd Zain et al. 2012; Pakdeenarong et al. 2014; Ribas et al. 2016; Chaisiri et al. 2017). Previous studies have revealed that helminth species richness and prevalence of infection can vary substantially based on different factors, including host species identity, gender, and maturity, as well as geographical distribution and human-modified habitat (Chaisiri et al. 2012; Krasnov et al. 2012; Mohd Zain et al. 2012; Palmeirim et al. 2014). Regarding habitat and landuse, rodents living in undisturbed habitats, i.e. forest and forested areas, seem to harbor a greater helminth diversity than those living in habitats with high human activity, i.e. households and human built-up land (Chaisiri et al. 2012; Pakdeenarong et al. 2014; Palmeirim et al. 2014; Chaisiri et al. 2017). This information is important not only to assess the existence of potential zoonotic health threats for humans, but also for updating research in ecological parasitology, as parasitism is an important interaction component for ecosystem functioning. Healthy and balanced ecosystems are often those characterized as parasite-rich (Cable et al. 2017). Studies on parasite transmission dynamics through the estimation of parasite abundance and diversity are of importance to monitor the links between biodiversity and health in fast changing environments.

In this chapter, we investigate a diverse assemblage of gastrointestinal helminths in the *Rattus rattus*- complex collected from the Southeast Asian mainland countries: Thailand, Lao PDR, and Cambodia. Using ecological analysis approaches, we assess helminth diversity and patterns of species assemblage across stratified human-dominated habitats. In addition, association patterns among helminth species are investigated using a probabilistic model of species co-occurrence.

## 2.2 Helminth Diversity in the *Rattus rattus*- Complex Based on Geographical Distribution and Stratified Habitats

A dataset on rodents and their helminths was extracted from several research projects conducted in Southeast Asia (from 2008 to present): community ecology of rodents and their pathogens in a changing environment (CERoPath: ANR07 BDIV012); local impacts and perceptions of global changes: biodiversity, health and zoonoses in South-East Asia (BioDivHealthSEA: ANR 11 CPEL 002); predictive scenarios of health in Southeast Asia: linking land use and climate changes to infectious diseases (FutureHealthSEA: ANR 17 CE35 0003 02); and potentially zoonotic infectious diseases at animal–human interface in Bangkok metropolitan: extensive investigation in Urban Public Park (MRG6180023). Data on helminth infections in murid rodents were collected from 22 study sites in 3 countries: Thailand (14 sites; Bangkok, Buriram, Chantaburi, Chiang Rai, Kalasin, Kanchanaburi, Loei, Nakhon Ratchasima, Nan, Phayao, Prachuab Khiri Khan, Songkhla, Tak, and Udon Thani); Lao PDR (4 sites; Champasak, Houaphan, Luang Prabang, and Vientiane); and Cambodia (4 sites: Mondolkiri, Pursat, Sihanouk, and Steung Treng) (Fig. 2.1). In brief, at each study site, rodent trapping was divided equally into three different

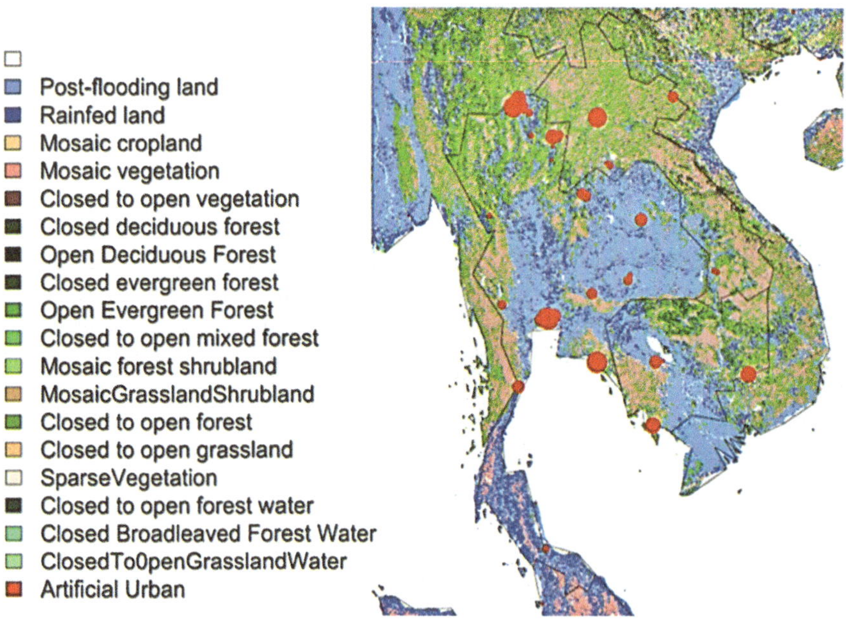


**Fig. 2.1** Distribution of rodent sampling sites across the three mainland Southeast Asian countries: Thailand, Lao PDR, and Cambodia. Pixel colors on the map represent associated land cover labels. The size of red circles indicates helminth species richness corresponded to each site

types of habitats with respect to human landuse (anthropization index), spanning low to high levels of disturbance, i.e. forest, peridomestic area, and human settlement.

After rodent euthanization, the gastrointestinal tracts were isolated and examined for helminth infections (see Chaisiri et al. 2012, 2017; Pakdeenarong et al. 2014; Palmeirim et al. 2014; Paladsing et al. 2020). To control for host taxonomic effects, we selected only helminth data from the *R. rattus*-complex for the following analyses. A high diversity of gastrointestinal parasites (32 species) was recorded in the species complex ($n = 813$), including cestodes (5 species), trematodes (5 species), nematodes (21 species), acanthocephalans (1 species), and pentastomids (1 species). Nine possible zoonotic helminths were found with trematodes: *Echinostoma malayanum*, *E. revolutum* and *Plagiorchis muris*; cestodes: *Hymenolepis diminuta, H. nana* and *Raillietina celebensis*; nematodes: *Gongylonema neoplasticum* and *Cyclodontostomum purvisi;* and the acanthocephalan: *Moniliformis moniliformis*. These species have been reported to infect humans in this region and also worldwide (Bhaibulaya and Indrangarm 1975; Rougier et al. 1981; Hong et al. 1996; Berenji et al. 2007; Pasuralertsakul et al. 2008; Toledo and Esteban 2016; Panti-May et al. 2020).

Helminth diversity, observed helminth species richness (HSR), species richness estimators (1st ordered Jackknife and Chao1) and Shannon diversity index (H′) were estimated across study sites and habitat types using the "BiodiversityR" package (Kidnt and Coe 2005) implemented in R freeware (R Core Team 2020). Helminth diversity in the *R. rattus*-complex differed with respect to geographical distribution and habitat type (Table 2.1). Rodents from several locations, e.g. Chiang Rai, Chantaburi, Luang Prabang, Nan, Mondolkiri and even public parks in a big city like Bangkok harbored a great diversity of helminth species. There was no significant association between helminth species richness and latitudinal gradient (Spearman correlation $R = 0.466$, $t = 1.826$, $p = 0.092$; see also Fig. 2.1). Unlike free-living organisms, much of the current relationship between parasite diversity and latitudinal gradient is unclear (Poulin and Morand 2000). However, Preisser (2019) revealed in their meta-analysis a trend to high helminth species richness (particularly nematode species richness) with lower latitude in cricetid rodents.

In terms of human-dominated habitats, *R. rattus*-complex species trapped from peridomestic habitat exhibited a greater helminth diversity (Jack1 = 40.97, Chao = 64.91 and H′ = 1.58) than those living in human settlements (Jack1 = 25.96, Chao = 27.13 and H′ = 1.54) and forest areas (Jack1 = 13.98, Chao = 13.98 and H′ = 1.42). Helminth species accumulation curves illustrate differences in parasite species richness among categorized habitats (Fig. 2.2). Peridomestic areas, as a transitional habitat (in a sense of ecotone) between forest and human settlement, appeared to be a hotspot for helminth species richness. Such an agricultural resource-rich landscape in a peridomestic area, e.g. fragmented habitats with field crops, plantations, orchards, and fallows, potentially provide favorable conditions for helminth transmission (Froeschke and Matthee 2014). Besides, Morand and Bordes (2015) pointed out that apart from other intrinsic (e.g., host attributes, investment in immune defenses, population density, and geographical range) and extrinsic determinants (e.g., landuse characteristics, meteorological conditions, and season), host

**Table 2.1** Diversity observations and estimation of gastrointestinal helminth infection in the *Rattus rattus*-complex from mainland Southeast Asia: comparison among study sites and habitat types

| Group | *n* | HSR | Jack1 [95% CI] | Chao [95% CI] | $H'$ [95% CI] |
|---|---|---|---|---|---|
| *Site* | | | | | |
| Bangkok (THA) | 132 | 14 | 17.96 [17.07–18.84] | 19.9 [18.6–21.19] | 1.58 [1.5–1.65] |
| Buriram (THA) | 35 | 5 | 5 [3.28–6.72] | 5 [2.48–7.52] | 0.94 [0.78–1.09] |
| Champasak (LAO) | 13[a] | 4 | 4.92 [2.09–7.74] | 4.15 [0.01–8.28] | 1.16 [0.91–1.41] |
| Chanthaburi (THA) | 61 | 12 | 17.9 [16.59–19.2] | 29.7 [27.79–31.61] | 1.04 [0.92–1.15] |
| Chiang Rai (THA) | 57 | 14 | 18.91 [17.56–20.25] | 20.14 [18.16–22.11] | 0.88 [0.76–0.99] |
| Houaphan (LAO) | 13[a] | 6 | 8.76 [5.93–11.58] | 8.76 [4.62–12.89] | 0.56 [0.31–0.81] |
| Kalasin (THA) | 33 | 7 | 7 [5.23–8.77] | 7 [4.41–9.59] | 1.45 [1.29–1.61] |
| Kanchanaburi (THA) | 23 | 5 | 6.91 [4.78–9.03] | 6.91 [3.81–10.01] | 1.35 [1.16–1.53] |
| Loei (THA) | 26 | 8 | 9.92 [7.92–11.91] | 8.96 [6.03–11.88] | 1.06 [0.88–1.23] |
| Luang Prabang (LAO) | 82 | 12 | 16.93 [15.80–18.05] | 24.34 [22.69–25.98] | 1.07 [0.97–1.17] |
| Mondolkiri (CAM) | 46 | 9 | 12.91 [11.40–14.41] | 14.86 [12.66–17.05] | 0.92 [0.78–1.05] |
| Nakhon Ratchasima | 3[a] | 6 | 8.66 [2.77–14.54] | 11.33 [2.72–19.93] | 0.89 [0.36–1.41] |
| Nan (THA) | 56 | 10 | 13.92 [12.55–15.28] | 15.89 [13.89–17.88] | 1.42 [1.29–1.54] |
| Prachuap Khiri Khan (THA) | 35 | 7 | 9.91 [8.18–11.63] | 9.91 [7.39–12.42] | 1.48 [1.32–1.63] |
| Pursat (CAM) | 33 | 7 | 8.93 [7.15–10.70] | 8.93 [6.33–11.52] | 1.34 [1.18–1.49] |
| Sihanouk (CAM) | 37 | 9 | 9.97 [8.29–11.64] | 9.24 [6.79–11.68] | 1.73 [1.58–1.87] |
| Songkla (THA) | 53 | 4 | 4 [2.6–5.4] | 4 [1.95–6.05] | 1.38 [1.25–1.51] |
| Steung Treng (CAM) | 6[a] | 0 | NA | NA | NA |
| Tak (THA) | 9[a] | 3 | 3.88 [0.48–7.27] | 3.44 [0–8.41] | 0.76 [0.23–1.28] |
| Udon Thani (THA) | 1[a] | 1 | 1 [0–11.19] | 1 [0–15.9] | NA |
| Vientiane (LAO) | 8[a] | 5 | 6.75 [3.14–10.35] | 5.87 [0.61–11.13] | 1.31 [0.99–1.62] |
| *Habitat type* | | | | | |
| Forest | 104 | 12 | 13.98 [11.38–16.57] | 13.98 [8.91–19.05] | 1.42 [1.41–1.43] |

(continued)

**Table 2.1** (continued)

| Group | *n* | HSR | Jack1 [95% CI] | Chao [95% CI] | *H′* [95% CI] |
|---|---|---|---|---|---|
| Peridomestic | 431 | 29 | 40.97 [39.69–42.24] | 64.91 [62.41–67.4] | 1.58 [1.57–1.58] |
| Settlement | 227 | 19 | 25.96 [24.2–27.71] | 27.13 [23.69–30.56] | 1.54 [1.53–1.55] |

Abbreviation: *N* number of host examined, *HSR* observed helminth species richness, *Jack1* first-ordered Jackknife, *Chao* Chao richness estimator and *H′* Shannon diversity index
[a]Indicate the groups with too low a sample size that were excluded from subsequent analyses

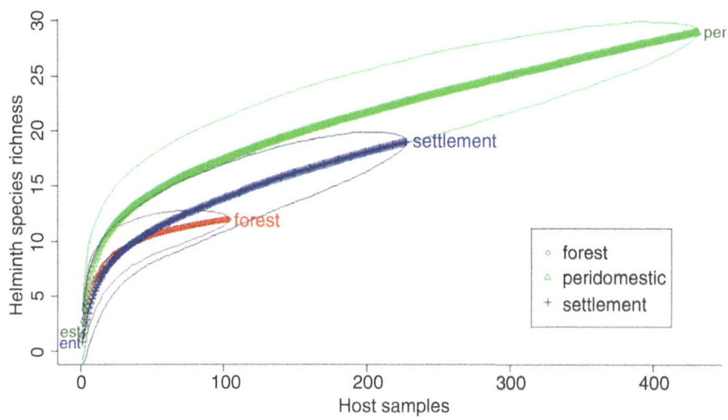


**Fig. 2.2** Helminth species accumulation curves among different habitats. These represent the mean values with their standard deviations from random permutations of the data

habitat specialization (behavioral trait) also explained the variation in parasite species richness. In mainland Southeast Asia, the *R. rattus*-complex (i.e., *R. tanezumi* and *Rattus* species phylogenetic R3) has been well documented as habitat generalists (low habitat preference) occurring in all types of habitats (Palmeirim et al. 2014; Blasdell et al. 2015). Host species that forage in more diverse habitats tend to be exposed to a greater number and diversity of parasites than habitat specialist species. Moreover, roaming in a variety of habitats could increase contact rates with other host species, resulting in an increasing chance for parasite sharing particularly generalist parasite species (low host specificity) (Morand and Bordes 2015).

Host-parasite network analysis (bipartite) was conducted to explore the interactions and patterns of helminth species assemblage among different habitat types. This was performed using "vegan" (Oksanen et al. 2013) and "bipartite" packages (Dormann et al. 2009) in R freeware. Network modularity, a measure of clustering (modular) structure within the host–parasite network was also computed using the "computeModules" function and illustrated graphically using the "plotModuleWeb" function. The higher the modularity, the more sub-communities are dependently clustered in the network (Dormann et al. 2009; Fortuna et al. 2010). Host–parasite

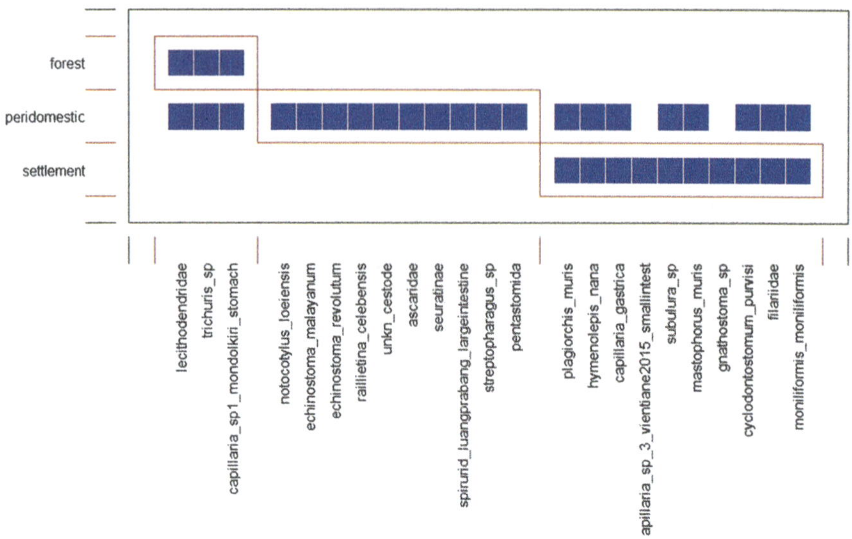


**Fig. 2.3** Composite panel of identified modules based on bipartite network analysis of helminth species assemblages among different habitat types

association at the whole community level (all study sites) was explored among the three habitats and 32 helminth species through a bipartite plot and assessment of network modularity (Fig. 2.3). Again, we found a similar trend; the majority of parasite species occurred in the *R. rattus*-complex from peridomestic areas. The analysis identified three modules (or subgroups) of host–parasite association pattern (network modularity = 0.325) with Group 1 that includes three helminths species in forests and peridomestic habitats, potentially involving in sylvatic cycle; Group 2 with ten helminth species strictly found in peridomestic habitat; and Group 3 with ten species recorded in peridomestic and human settlement habitats, showing adaptation to anthropogenic activities. In addition, there were another ten helminth species: *Raillietina* sp., *H. diminuta*, *Heterakis spumosa*, *Syphacia muris*, *Physaloptera ngoci*, *Protospirura siamensis*, *Pterygodermatites* sp., *G. neoplasticum*, and Trichostrongylidae sp. found in all habitat types (habitat generalists). These ten species were ignored in the network modularity analysis as they were present everywhere, and they are not shown in Fig. 2.3.

## 2.3 Helminth Species Co-occurrences in the *Rattus rattus*-Complex

Helminth species co-occurrence from the host–parasite dataset was investigated using a probabilistic model of species co-occurrence implemented in the "co-occur" package (Griffith et al. 2016), R freeware (R Core Team 2020). The

**Table 2.2** List of 12 selected helminths in the *Rattus rattus*-complex from the Southeast Asian mainland with the number of species pairs to classified co-occurrence patterns (positive, negative, or random associations)

| Helminth species | Taxonomic group | Target organ | Number of species co-occurrence patterns (pairs to the others) | | |
|---|---|---|---|---|---|
| | | | Positive | Negative | Random |
| *Raillietina sp.* | Cestoda: Davaineidae | SI | 3 | 4 | 4 |
| *Raillietina celebensis* | Cestoda: Davaineidae | SI | 2 | 1 | 8 |
| *Hymenolepis diminuta* | Cestode: Hymenolepididae | SI | 3 | 1 | 7 |
| *Capillaria gastrica* | Nematoda: Capillariidae | ST | 5 | 0 | 6 |
| *Capillaria* sp. | Nematoda: Capillariidae | ST | 2 | 0 | 9 |
| *Protospirura siamensis* | Nematoda: Spiruridae | ST | 1 | 1 | 9 |
| *Physaloptera ngoci* | Nematoda: Physalopteridae | ST | 1 | 2 | 8 |
| *Gongylonema neoplasticum* | Nematoda: Gongylonematidae | ST | 4 | 2 | 5 |
| *Pterygodermatites* sp. | Nematoda: Rictulariidae | SI | 6 | 3 | 2 |
| Trichostrongylidae | Nematoda: Trichostrongylidae | SI | 5 | 4 | 2 |
| *Heterakis spumosa* | Nematoda: Heterakidae | LI | 2 | 1 | 8 |
| *Syphacia muris* | Nematoda: Oxyuridae | LI | 5 | 0 | 6 |

*ST* stomach, *SI* small intestine, *LI* large intestine

probabilistic model of species co-occurrence allows us to obtain the probability that paired species in a dataset co-occur at frequencies either lower or higher than the observed frequencies of co-occurrence (Veech 2013). In the sense of pairwise co-occurrence patterns, the analysis classifies paired-association patterns into three categories: positive, negative, or random parasite–parasite associations based on a preferable significance level. Positive association occurs, if one parasite mutually enhances the presence and transmission of another parasite species (Johnson and Hoverman 2012; Lass et al. 2012; Dallas et al. 2019). In contrast, direct competition between parasite species for host resources and internal niche space, as well as through an indirect alteration of host immune regulation against another species, would generally lead to a negative association (Holland 1984; Sousa 1992; Frontera et al. 2005; Griffith et al. 2016).

Here, helminth species co-occurrence analysis was performed at a global scale (all sites), a local scale (by study sites), as well as at the levels of habitat type and host gender to identify patterns of species associations. For the global scale analysis, the probabilistic model revealed significant association patterns (positive and negative associations) among the 12 selected helminth species (Table 2.2 and Fig. 2.4). The model was applied to those paired species whose expected number of co-occurrences was greater than 1.0 with a significance level at 95% to classify species associations. Of the 528 species pair combinations, 457 pairs (86.55%) were removed from the

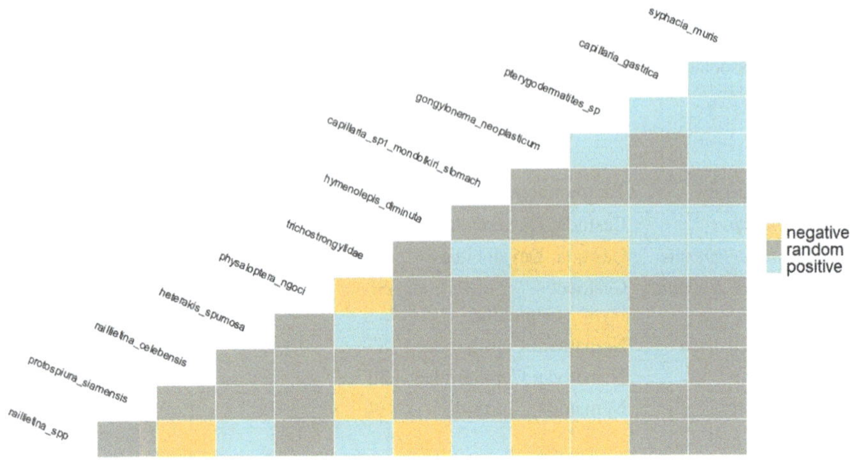


**Fig. 2.4** A heatmap visualization of the paired helminth species revealed by a species co-occurrence analysis. Blue, orange, and gray labels indicate significant positive-, negative-, and random association patterns, respectively

analysis because the expected co-occurrence was lower than 1.0, and 71 pairs were retained for analysis. The nematode *Pterygodermatites* sp. was the species with the greatest number of positive associations (6 pairs), followed by *Capillaria gastrica* (5 pairs), *Syphacia muris* (5 pairs), and Trichostrongylidae (5 pairs), whereas the parasite species with the most negative associations (4 pairs) was Trichostrongylidae. *Raillietina* sp. and *Capillaria* sp. in the stomach, while *Protospirura siamensis* showed mostly random associations (9 pairs for each species), followed by *Raillietina celebensis* (8 pairs), *Physaloptera ngoci* (8 pairs), and *Heterakis spumosa* (8 pairs). Positive associations (20 species pairs) were more common than negative associations (9 species pairs). This is in accordance to a general trend from previous studies suggesting that positive associations are more common than negative associations in the helminth infracommunities of small mammals, i.e. bats and rodents (Lotz and Font 1991; Behnke et al. 2005; Dallas et al. 2019), or birds (Forbes et al. 1999). Several factors may explain how some parasite species facilitate colonization by another species when the environmental conditions are suitable for transmission and establishment in either definitive or intermediate hosts. In addition, immune-mediated interactions among hosts and parasites can play an important role. In the case of successive infections, the first parasite species may induce immuno-suppression, facilitating the establishment of subsequent infections by other parasite species (Graham 2008). However, in contrast, immunity triggered by an earlier parasitic infection may result in the subsequent inhibition of another parasite infection, particularly against microparasites (Jolles et al. 2008; Salgame et al. 2013). In addition to these, differences in helminth species association patterns are potentially influenced by other extrinsic or intrinsic factors, e.g. host maturity, gender, behavior, and exposure history, as well as habitat

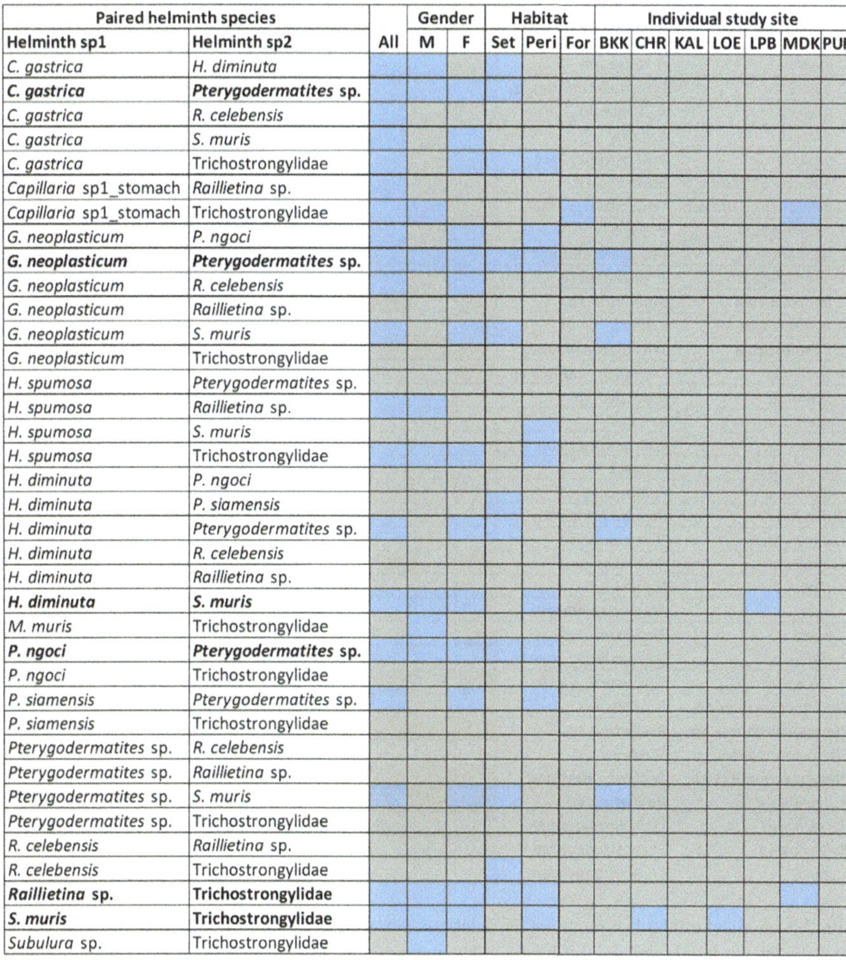


**Fig. 2.5** A checkerboard of positive associations among paired helminth species across different analysis levels: whole dataset, by host gender, habitat, and study site. Significant associations are labeled in blue. The paired species with prominent associations are highlighted in bold. *M* male, *F* female, *Set* settlement, *Peri* peridomestic, *For* forest, *BKK* Bangkok, *CHR* Chiang Rai, *KAL* Kalasin, *LOE* Loei, *LPB* Luang Prabang, *MDK* Mondolkiri, *PUR* Pursat

and seasonal variation (Haukisalmi and Henttonen 1993; Forbes et al. 1999; Behnke 2008; Telfer et al. 2008; Johnson and Buller 2011).

Investigation at different levels (whole dataset, by host gender, habitat type, and study site) revealed paired helminth species with significant positive associations (facilitation), e.g. *G. neoplasticum*, *Pterygodermatites* sp., *H. diminuta*, *S. muris*, *P. ngoci*, *Pterygodermatites* sp., *Raillietina* sp., Trichostrongylidae, and *S. muris*—Trichostrongylidae. In contrast, pairs with a significant negative association (competition) included *Pterygodermatites* sp.—Trichostrongylidae, *G. neoplasticum*—

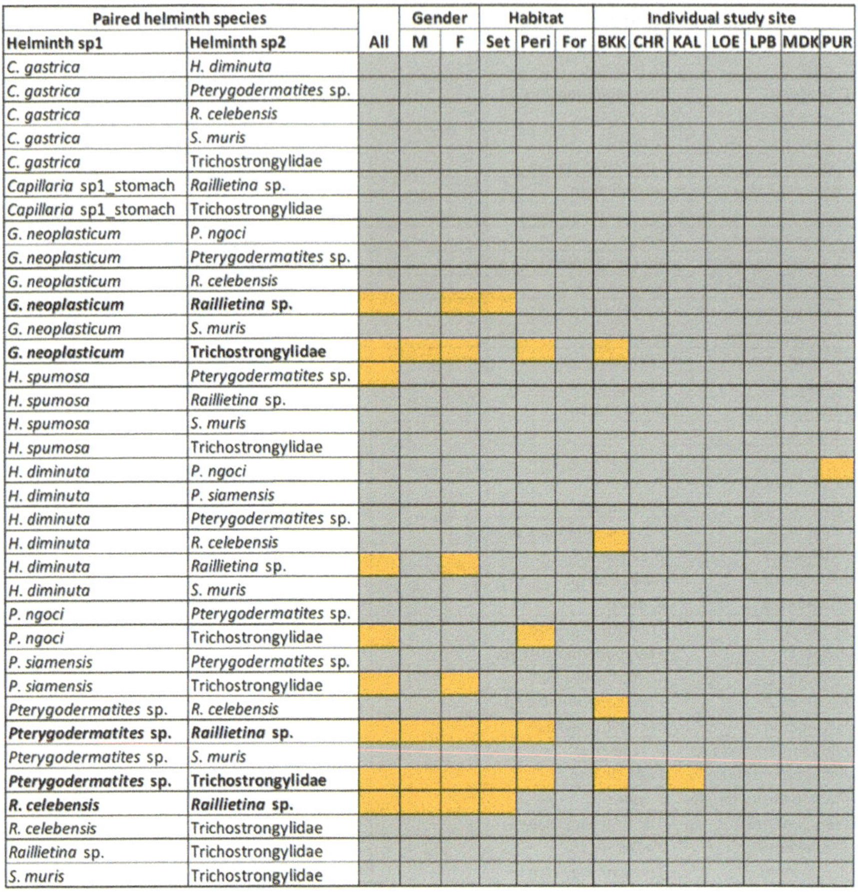


**Fig. 2.6** A checkerboard of negative associations among paired helminth species across different analysis levels: whole dataset, by host gender, habitat, and study site. Significant associations are labeled in orange. The paired species with prominent associations are highlighted in bold. *M* male, *F* female, *Set* settlement, *Peri* peridomestic, *For* forest, *BKK* Bangkok, *CHR* Chiang Rai, *KAL* Kalasin, *LOE* Loei, *LPB* Luang Prabang, *MDK* Mondolkiri, *PUR* Pursat

Trichostrongylidae, and *Pterygodermatites* sp.—*Raillietina* sp. (see more details in Figs. 2.5 and 2.6). Interestingly, the two pairs of negative association: *Pterygodermatites* sp.—Trichostrongylidae and *Pterygodermatites* sp.—*Raillietina* sp. shared the same microhabitat in the hosts; all of them parasitize the small intestine. Almost all positive association pairs were the helminths infecting different target organs, with the only exception being *Raillietina* sp.—Trichostrongylidae (both were found in the small intestine). Negative association was likely to occur when the two parasites shared the same host tissue, probably because they compete for similar host resources and/or for space. This observation is in line with a recent study of parasite communities in small mammals from the Sonoran Desert, New

Mexico (Dallas et al. 2019). Our findings are solely based on the analysis of a probabilistic model. Controlled infection studies should be performed for investigating potential explanatory factors and mechanisms.

The results also indicated that trichostrongylid nematodes, Clade V: Strongylida (Durette-Desset et al. 1999) tended to compete (negative association) with spirurid nematodes Clade III: Spirurida (Nadler et al. 2007), *G. neoplasticum* and *Pterygodermatites* sp., and there is no positive association shown between the two nematode taxa. Similar observations were reported in previous studies: antagonistic interactions between *Metastrongylus apri*, Clade V: Strongylida (Carreno and Nadler 2003) and *Ascaris suum*, Clade III: Ascaridida (Nadler and Hudspeth 2000) in experimental pigs (Frontera et al. 2005). Negative associations between parasites from different clades, which suggest taxonomic and evolutionary influences, should be explored.

## 2.4 Conclusions

Our understanding of host–parasite diversity in Southeast Asia remains rudimentary. Here we examine the *Rattus rattus* species complex to explore diversity and species association patterns of helminth infections in relation to the effect of human-dominated habitats in mainland Southeast Asian countries. The analyses presented here show that this synanthropic species complex maintains a high diversity of helminth species in peridomestic habitats and acts as a bridge between forested areas and human settlements. A deeper understanding of associated factors favoring parasite colonization, diversity and species association patterns is needed at the interface between hosts and their habitats in fast changing environments in Southeast Asia. The methods used in this chapter are a way to improve our knowledge, but call for further studies in the fields of parasite ecology and disease ecology. We propose, as a future perspective, to extend the analyses to a larger scale, e.g. all rodent species endemic in the region, exploring other ecological factors influencing parasite diversity and uncovering mechanisms of parasite co-occurrences in the host population while taking into account the evolutionary biology of this diversified group. In addition, the study on parasite association patterns can be extended to other co-occurring organisms within the same host such as the microbiome and virome.

## References

Aplin K, Brown P, Jacob J, Krebs C, Singleton G (2003) Field methods for rodent studies in Asia and the Indo-Pacific. ACIAR Monograph No. 100. Australian Centre for International Agricultural Research, Canberra

Aplin KP, Suzuki H, Chinen AA, Chesser RT, Ten Have J, Donnellan SC, Austin J, Frost A, Gonzalez JP, Herbreteau V, Catzeflis F, Soubrier J, Fang YP, Robins J, Matisoo-Smith E, Bastos AD, Maryanto I, Sinaga MH, Denys C, Van Den Bussche RA, Conroy C, Rowe K,

Cooper A (2011) Multiple geographic origins of commensalism and complex dispersal history of black rats. PLoS One 6:e26357. https://doi.org/10.1371/journal.pone.0026357
Behnke JM (2008) Structure in parasite component communities in wild rodents: predictability, stability, associations and interactions or pure randomness? Parasitology 135:751–766
Behnke JM, Gilbert FS, Abu-Madi MA, Lewis JW (2005) Do the helminth parasites of wood mice interact? J Anim Ecol 74:982–993
Berenji F, Fata A, Hosseininejad ZA (2007) Case of *Moniliformis moniliformis* (Acanthocephala) infection in Iran. Korean J Parasitol 45:145–148
Bhaibulaya M, Indrangarm S (1975) Man, an accidental host of *Cyclodontostomum purvisi* (Adams, 1933), and the occurrence in rats in Thailand. Southeast Asian J Trop Med Public Health 6:391–394
Blasdell K, Bordes F, Chaisiri K, Chaval Y, Claude J, Cosson JF, Latinne A, Michaux J, Morand S, Pagès M, Tran A (2015) Progress on research on rodents and rodent-borne zoonoses in Southeast Asia. Wildl Res 42:98–107
Burgin CJ, Colella JP, Kahn PL, Upham NS (2018) How many species of mammals are there? J Mammal 99:1–14
Cable J, Barber I, Boag B, Ellison AR, Morgan ER, Murray K, Pascoe EL, Sait SM, Wilson AJ, Booth M (2017) Global change, parasite transmission and disease control: lessons from ecology. Philos Trans R Soc B 372:20160088. https://doi.org/10.1098/rstb.2016.0088
Carreno R, Nadler SA (2003) Phylogenetic analysis of the Metastrongyloidea (Nematoda: Strongylida) inferred from ribosomal RNA gene sequences. J Parasitol 89:965–973
Centers for Disease Control and Prevention (2017) Diseases directly transmitted by rodents. In: CDC Rodents homepage. https://www.cdc.gov/rodents/diseases/direct.html. Accessed 15 Aug 2020
Chaisiri K, Chaeychomsri W, Siruntawineti J, Ribas A, Herbreteau V, Morand S (2012) Diversity of gastrointestinal helminthes among murid rodents from northern and northeastern Thailand. Southeast Asian J Trop Med Public Health 43:21–28
Chaisiri K, Siribat P, Ribas A, Morand S (2015) Potentially zoonotic helminthiases of murid rodents from the Indo-Chinese Peninsula: impact of habitat and the risk of human infection. Vector Borne Zoonotic Dis 15:73–85
Chaisiri K, Chou M, Siew CC, Morand S, Ribas A (2017) Gastrointestinal helminth fauna of rodents from Cambodia: emphasizing the community ecology of host–parasite associations. J Helminthol 91:726–738
Claveria FG, Causapin J, Guzman MA, Toledo MG, Salibay C (2005) Parasite biodiversity in *Rattus spp.* caught in wet markets. Southeast Asian J Trop Med Public Health 36:146–148
Dallas TA, Laine AL, Ovaskainen O (2019) Detecting parasite associations within multi-species host and parasite communities. Proc Biol Sci 286:20191109. https://doi.org/10.1098/rspb.2019.1109
Dormann CF, Fruend J, Bluethgen N, Gruber B (2009) Indices, graphs and null models: analyzing bipartite ecological networks. Open Ecol J 2:7–24
Durette-Desset MC, Hugot JP, Darlu P, Chabaud AG (1999) A cladistic analysis of the Trichostrongyloidea (Nematoda). Int J Parasitol 29:1065–1086
Fabre P, Hautier L, Dimitrov D, Douzery DJP (2012) A glimpse on the pattern of rodent diversification: a phylogenetic approach. BMC Evol Biol 12:88. https://doi.org/10.1186/1471-2148-12-88
Forbes MR, Alisauskas RT, McLaughlin JD, Cuddington KM (1999) Explaining co-occurrence among helminth species of lesser snow geese (*Chen caerulescens*) during their winter and spring migration. Oecologia 120:613–620
Fortuna MA, Stouffer DB, Olesen JM, Jordano P, Mouillot D, Krasnov BR, Poulin R, Bascompte J (2010) Nestedness versus modularity in ecological networks: two sides of the same coin? J Anim Ecol 79:811–817

Froeschke G, Matthee S (2014) Landscape characteristics influence helminth infestations in a peri-domestic rodent - implications for possible zoonotic disease. Parasit Vectors 7:393. https://doi.org/10.1186/1756-3305-7-393

Frontera E, AlcaideaJ M, Domínguez-Alpízar L, Boes J, Reina D, Navarrete I (2005) Evidence of interaction between *Ascaris suum* and *Metastrongylus apri* in experimentally infected pigs. Vet Parasitol 127:295–301

Graham AL (2008) Ecological rules governing helminth–microparasite coinfection. Proc Natl Acad Sci 105:566–570

Griffith DM, Veech JA, Marsh CJ (2016) Cooccur: probabilistic species co-occurrence analysis in R. J Stat Softw 69(2):1–17. https://doi.org/10.18637/jss.v069.c02

Haukisalmi Y, Henttonen H (1993) Coexistence in helminths of the bank vole, *Clethrionomys glareolus*. I. Patterns of co-occurrence. J Anim Ecol 62:221–229

Holland CV (1984) Interactions between *Moniliformis* (Acanthocephala) and *Nippostrongylus* (Nematoda) in the small intestine of laboratory rats. Parasitology 88:303–315

Hong SJ, Woo HC, Chai JY (1996) A human case of *Plagiorchis muris* (Tanabe, 1922: Digenea) infection in the Republic of Korea: freshwater fish as a possible source of infection. J Parasitol 82:647–649

Jarvi SI, Quarta S, Jacquier J, Howe K, Bicakci D, Dasalla C, Lovesy N, Snook K, McHugh R, Niebuhr CN (2017) High prevalence of *Angiostrongylus cantonensis* (rat lungworm) on eastern Hawaii Island: a closer look at life cycle traits and patterns of infection in wild rats (*Rattus spp.*). PLoS ONE 12:e0189458. https://doi.org/10.1371/journal.pone.0189458

Johnson PT, Buller ID (2011) Parasite competition hidden by correlated coinfection: using surveys and experiments to understand parasite interactions. Ecology 92:535–541

Johnson PT, Hoverman JT (2012) Parasite diversity and coinfection determine pathogen infection success and host fitness. Proc Natl Acad Sci 109:9006–9011. https://doi.org/10.1073/pnas.1201790109

Jolles AE, Ezenwa VO, Etienne RS, Turner WC, Olff H (2008) Interactions between macroparasites and microparasites drive infection patterns in free-ranging African buffalo. Ecology 89:2239–2250

Kidnt R, Coe R (2005) Tree diversity analysis: a manual and software for common statistical methods for ecological and biodiversity studies. World Agroforestry Centre (ICRAF), Nairobi

Krasnov BR, Bordes F, Khokhlova IS, Morand S (2012) Gender-biased parasitism in small mammals: patterns, mechanisms, consequences. Mammalia 76:1–13

Lass S, Hudson PJ, Thakar J, Harvill E, Albert R, Perkins SE (2012) Generating super-shedders: co-infection increases bacterial load and egg production of a gastrointestinal helminth. J R Soc Interface 10:20120588. https://doi.org/10.1098/rsif.2012.0588

Lotz JM, Font WF (1991) The role of positive and negative interspecific associations in the organization of communities of intestinal helminths of bats. Parasitology 103(1):127–138

McGarry JW, Higgins A, White NG, Pounder KC, Hetzel U (2015) Zoonotic helminths of urban brown rats (*Rattus norvegicus*) in the UK: neglected public health considerations? Zoonoses Public Health 62:44–52

Meerburg BG, Singleton GR, Kijlstra A (2009) Rodent-borne diseases and their risks for public health. Crit Rev Microbiol 35:221–270

Mohd Zain SN, Behnke JM, Lewis JW (2012) Helminth communities from two urban rat populations in Kuala Lumpur, Malaysia. Parasit Vectors 5:47. https://doi.org/10.1186/1756-3305-5-47

Morand S, Bordes F (2015) Parasite diversity of disease-bearing rodents of Southeast Asia: habitat determinants and effects on sexual size dimorphism and life-traits. Front Ecol Evol 3:110. https://doi.org/10.3389/fevo.2015.00110

Morand S, Krasnov BR, Poulin R (2006) Micromammals and macroparasites: from evolutionary ecology to management. Springer, Tokyo, p 647

Morand S, Bordes F, Chen HW, Claude J, Cosson JF, Galan M, Czirják GÁ, Greenwood AD, Latinne A, Michaux J, Ribas A (2015) Global parasite and *Rattus* rodent invasions: the consequences for rodent borne diseases. Integ Zool 10:409–423

Nadler SA, Hudspeth DS (2000) Phylogeny of the Ascaridoidea (Nematoda: ASCARIDOIDEA) based on three genes and morphology: hypotheses of structural and sequence evolution. J Parasitol 86:380–393

Nadler SA, Carreno RA, Mejia-Madrid H, Ullberg J, Pagan C, Houston R, Hugot JP (2007) Molecular phylogeny of clade III nematodes reveals multiple origins of tissue parasitism. Parasitology 134:1421–1442

Oksanen J, Blanchet FG, Kindt R, Legendre P, Minchin PR, O'Hara RB, Simpson GL, Solymos P, Stevens MH, Wagner H (2013) Vegan: Community ecology package. R package version 2.0-10. Available at http://CRAN.R-project.org/package=vegan

Pagès M, Chaval Y, Herbreteau V, Waengsothorn S, Cosson JF, Hugot JP, Morand S, Michaux J (2010) Revisiting the taxonomy of the Rattini tribe: a phylogeny-based delimitation of species boundaries. BMC Evol Biol 10:184

Pakdeenarong N, Siribat P, Chaisiri K, Douangboupha B, Ribas A, Chaval Y, Herbreteau V, Morand S (2014) Helminth communities in murid rodents from southern and northern localities in Lao PDR: the role of habitat and season. J Helminthol 88:302–309

Paladsing Y, Boonsri K, Saesim W, Changsap B, Thaenkham U, Kosoltanapiwat N, Sonthayanon P, Ribas A, Morand S, Chaisiri K (2020) Helminth fauna of small mammals from public parks and urban areas in Bangkok Metropolitan with emphasis on community ecology of infection in synanthropic rodents. Parasitol Res. https://doi.org/10.1007/s00436-020-06897-9

Palmeirim M, Bordes F, Chaisiri K, Siribat P, Ribas A, Morand S (2014) Helminth parasite species richness in rodents from Southeast Asia: role of host species and habitat. Parasitol Res 113:3713–3726

Panti-May JA, Rodríguez-Vivas RI, García-Prieto L, Servián A, Costa F (2020) Worldwide over view of human infections with *Hymenolepis diminuta*. Parasitol Res 119:1997–2004

Paramasvaran S, Krishnasamy M, Lee HL, John J, Lokman H, Naseem BM, Rehana AS, Santhana RL (2005) Helminth infections in small mammals from Ulu Gombak Forest Reserve and the risk to human health. Trop Biomed 22:191–194

Pasuralertsakul S, Yaicharoen R, Sripochang S (2008) Spurious human infection with *Gongylonema*: nine cases reported from Thailand. Ann Trop Med Parasitol 102:455–457

Pham XD, Tran CL, Hasegawa H (2001) Helminths collected from Rattus spp. in Bac Ninh Province, Vietnam. Comp Parasitol 68:261–264

Poulin R, Morand S (2000) The diversity of parasites. Q Rev Biol 75:277–293

Preisser W (2019) Latitudinal gradients of parasite richness: a review and new insights from helminths of cricetid rodents. Ecography 42:1315–1330

R Core Team (2020) R: A language and environment for statistical computing. R Foundation for Statistical Computing, Vienna

Ranjbar MJ, Sarkari B, Mowlavi GR, Seifollahi Z, Moshfe A, Abdolahi KS, Mobedi I (2017) Helminth infections of rodents and their zoonotic importance in Boyer-Ahmad District, Southwestern Iran. Iran J Parasitol 12:572–579

Ribas A, Saijuntha W, Agatsuma T, Lamsan K, Poonlaphdecha S (2016) Helminths in rodents from wet markets in Thailand. Helminthologia 53:326–330

Rougier Y, Legros F, Durand JP, Cordoliani Y (1981) Four cases of parasitic infection by *Raillietina (R.) celebensis* (Kanicki, 1902) in French Polynesia. Trans R Soc Trop Med Hyg 75:121

Salgame P, Yap GS, Gause WC (2013) Effect of helminth-induced immunity on infections with microbial pathogens. Nat Immunol 14:1118–1126

Sousa WP (1992) Interspecific interactions among larval trematode parasites of freshwater and marine snails. Am Zool 32:583–592

Stuart AM, Singleton GR, Prescott CV (2015) Population ecology of the Asian house rat (*Rattus tanezumi*) in complex lowland agroecosystems in the Philippines. Wildl Res 42:165–175

Telfer S, Birtles R, Bennett M, Lambin X, Paterson S, Begon M (2008) Parasite interactions in natural populations: insights from longitudinal data. Parasitology 135:767–781

Toledo R, Esteban JG (2016) An update on human echinostomiasis. Trans R Soc Trop Med Hyg 110:37–45

Veech JA (2013) A probabilistic model for analysing species co-occurrence. Glob Ecol Biogeogr 22:252–260

Wilson D, Reeder D (2005) Mammal species of the world, 3rd edn. Johns Hopkins University Press, Baltimore

Wiroreno W (1978) Nematode parasites of rats in West Java, Indonesia. Southeast Asian J Trop Med Public Health 9:520–525

# Chapter 3
# Intestinal Protozoa: Their Role as Human Pathogens and Zoonoses

**Kotchaphon Vaisusuk and Weerachai Saijuntha**

**Abstract** Intestinal protozoa are single-celled eukaryotic microorganisms with nearly 10,000 species being parasitic in many invertebrates and most vertebrates. Intestinal protozoan infections are among the most common worldwide and contribute significantly to the burden of infectious diseases. While the morbidity and mortality caused by parasitic diseases affect people mainly in developing countries, protozoa also cause significant illness (e.g., outbreaks) in developed countries. These diseases are much more common in the tropics, including Southeast Asia, wherever sanitation is poor, making them a major health problem. Hence, it is crucial to have an enhanced understanding of the current status of the epidemiology of intestinal protozoan infections. In this review we provide an update on intestinal protozoan infections in Southeast Asian countries covering the diseases caused by *Giardia duodenalis*, *Entamoeba histolytica*, *Cryptosporidium,* and *Blastocystis* sp. Among them *Giardia duodenalis* is the best-known cause of protozoan gastrointestinal disease, producing significant but not life-threatening gastrointestinal distress and diarrhea. Dysentery, however, caused by *E. histolytica* is probably the most dangerous intestinal protozoan infection, although *Cryptosporidium* and *Blastocystis* sp. may cause diarrhea in healthy individuals and result in intractable, life-threatening illness in patients with acquired immunodeficiency syndrome or other immunosuppressive diseases.

**Keywords** Protozoa parasites · *Giardia* · *Cryptosporidium* · *Entamoeba* · *Blastocystis*

K. Vaisusuk
Department of Veterinary Technology and Veterinary Nursing, Faculty of Agricultural Technology, Rajabhat Maha Sarakham University, Maha Sarakham, Thailand

W. Saijuntha (✉)
Walai Rukhavej Botanical Research Institute, Biodiversity and Conservation Research Unit, Mahasarakham University, Maha Sarakham, Thailand
e-mail: weerachai.s@msu.ac.th



T. N. Petney et al. (eds.), *Biodiversity of Southeast Asian Parasites and Vectors causing Human Disease*, Parasitology Research Monographs 14,
https://doi.org/10.1007/978-3-030-71161-0_3

## 3.1 Introduction

Protozoan infections contribute significantly to the burden of gastrointestinal illness worldwide. Intestinal protozoa are single-celled eukaryotic microorganisms. There are more than 20,000 protozoan species of protozoa known and of these approximately 10,000 are parasites found in invertebrates and in almost all vertebrates. It has been estimated that as much as 60% of people worldwide are infected with intestinal parasites (World Health Organization (WHO) 1987). Intestinal protozoan infections are worldwide and are among the most common infection with 3.5 billion people being infected and some 450 million people are sick as a result of an infection (Haileeyesus and Beyene 2009). Protozoa have a rather simple and more primitive structure than other members of the animal kingdom. They contain membrane-bound nuclei and intra cellular organelles, and most protozoa have at least in some stages of their life cycle, structures such as flagella, cilia, or pseudopodia, which enable them to move, and, for some species, to obtain nutrients. The traditional classification of protozoa is based on their specific structures for movement (e.g. flagellates, ciliates, amoeba). Most protozoa are microscopic and typically live in moist conditions, and the majority are free-living. Protozoa may multiply asexually by binary or multiple fission, but some of them are capable of sexual reproduction too.

The morbidity and mortality due to parasitic diseases affect people mainly in developing countries. Protozoa, however, also cause significant illness in developed countries (Ortega et al. 2008). Several species of enteric protozoa are associated with diarrheal illnesses in humans, with some causing severe debilitating illness, especially in immunosuppressed people (Kucerova et al. 2011). These infections may cause anemia, malnutrition, and other physical and mental impairments especially in children. Symptoms of infection may include diarrhea, stomach pain, nausea or vomiting, bloating, liver abscesses, colitis, fatigue, etc. It is important to understand the epidemiology and appropriate prevention strategies for intestinal protozoan infections in order to limit and control the cause of the disease (Zeibig 1997). *Giardia*, *Cryptosporidium* spp., *Dientamoeba fragilis*, *Entamoeba* spp. (including non-pathogenic species), and *Cyclospora cayetanensis* are the most common pathogenic protozoa reported in developed settings (Fletcher et al. 2012). A parasite which is often referred to as a protozoan is *Blastocystis*, which is one of the most common intestinal parasites worldwide. For practical reasons, this organism has been included in this chapter on intestinal protozoan infections.

## 3.2 *Giardia duodenalis*

*Giardia duodenalis* (syn. *Giardia lamblia* and *Giardia intestinalis*) is a flagellated protozoon giving rise to giardiasis, one of the most common gastrointestinal infections of mammals, including people, with a worldwide distribution. The trophozoite of *G. duodenalis* is a pear-shaped cell, with a length, width, and thickness of 12–15, 6–8, and 1–2 μm, respectively (Fig. 3.1). Trophozoites contain two nuclei surrounded by nuclear envelopes and have a complex cytoskeleton, which maintains their shape and anchors the four pairs of flagella, the median body, and the ventral disk. The flagella are composed of microtubules in a typically eukaryotic 9 + 2 arrangement and are built from basal bodies located between the nuclei. The median bodies are formed by an irregular set of microtubules that have a comma-shaped structure, which varies in size and thickness, and are located transversally, perpendicular to the central axis. Cysts of *G. duodenalis* are oval in shape and range in size from 10 to 12 μm (range: 8 to 19 μm) (Garcia et al. 2018). These oval cysts are thick-walled with four nuclei and several internal fibers. The cyst wall is 0.3–0.5 μm in thickness and is formed by an outer filamentous layer and an inner membranous layer including two membranes that enclose the periplasmic space. The cyst wall is composed of carbohydrates in the form of N-acetylgalactosamine polymers and cyst wall proteins. The cytoplasm of the mature cyst contains four nuclei, the contracted flagella, and fragmented portions of the ventral disc (Fig. 3.1).

In contrast with *G. duodenalis* trophozoites, *G. duodenalis* cysts are infective. As few as 10 cysts can cause an infection (Rendtorff 1954). Cysts are ingested by consuming fecally contaminated food or water or by more direct fecal-oral self-transmission. They can survive outside the body for several months and are also relatively resistant to chlorination, UV exposure, and freezing. When cysts are ingested, the low pH of the stomach acid triggers excystation, in which phase the activated flagella breaks through the cyst wall. This occurs in the small intestine, specifically in the duodenum. Excystation releases trophozoites, with each cyst producing two trophozoites. Within the small intestine, the trophozoites reproduce asexually (longitudinal binary fission) and either float free in the lumen or attach to the intestinal mucosa. Trophozoites may encyst in the small intestine on their passage toward the colon. Encystation occurs most likely as a result of exposure to bile salts and/or fatty acids and is stimulated by an alkaline environment. It can also be triggered by water reabsorption in the colon (Bingham and Meyer 1979). Both

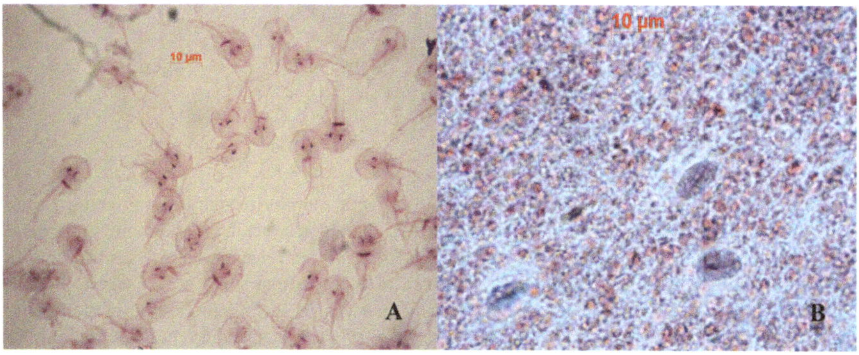


**Fig. 3.1** *Giardia duodenalis* trophozoites stained with Giemsa (**a**) and cysts stained with trichrome (**b**) (Photo by Kotchaphon Vaisusuk)

cysts and trophozoites are then passed in the feces, and the cysts are infectious immediately or shortly after being shed. Person-to-person transmission is possible. Many groups of animals are commonly infected, thus serving as a zoonotic reservoir (Fig. 3.2).

*Giardia duodenalis* is an important pathogen in both human and veterinary health. Evaluation of the extent of zoonotic transmission of the infection requires molecular characterization as there is considerable genetic variation within the

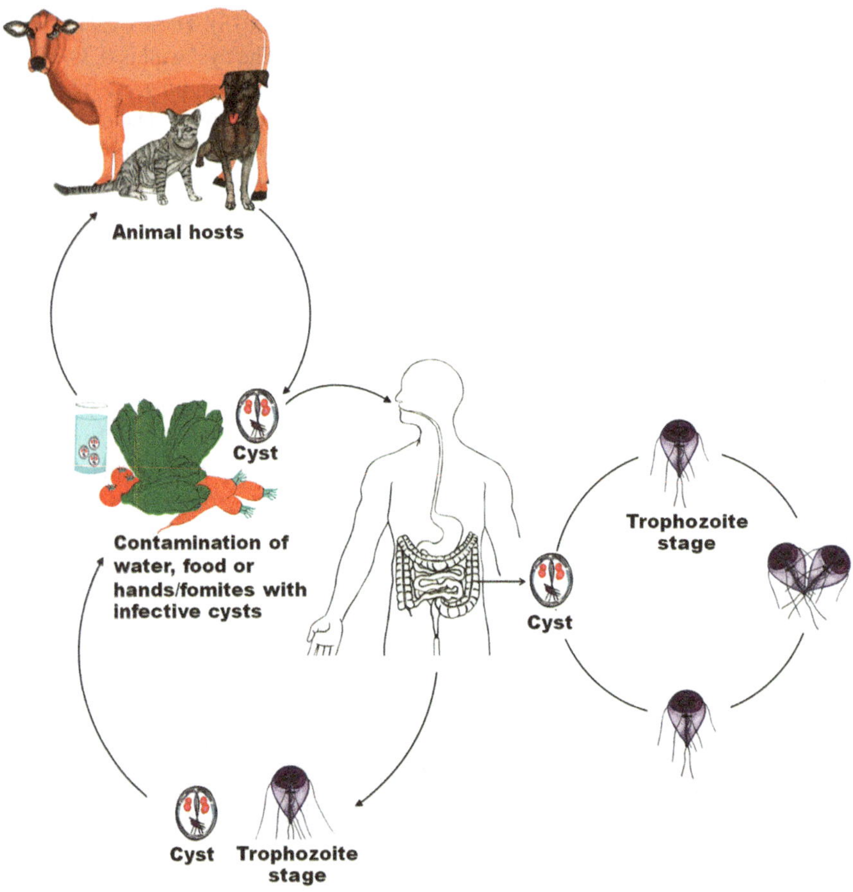


**Fig. 3.2** *Giardia* cysts are hardy and can survive for several months in cold water. Infection occurs by ingestion of cysts present in contaminated water, food, or by a direct fecal-oral route. In the small intestine, excystation results in the release of trophozoites. After release trophozoites multiply by longitudinal binary fission, remaining in the lumen of the proximal small bowel where they are found free in the lumen or attached to the mucosa by their ventral sucking discs. Encystation occurs as the parasites transit toward the colon. The cyst is the stage found most commonly in non-diarrheal feces. Because the cysts are infectious when passed in the stool or shortly afterwards, person-to-person transmission is possible. Many major animal groups can serve as a reservoir of zoonotic *Giardia* infections (Picture by Naruemon Bunchom)

species *G. duodenalis*. To date, eight major genetic groups (assemblages) have been identified, two of which (A and B) are found in both humans and non-human hosts, whereas the remaining six (C to H) are host-specific and do not infect humans (Feng and Xiao 2011; Cacciò and Lalle 2015). Assemblages C and D are canine-specific. However, assemblages A, B, and E have also been reported in dogs (Li et al. 2015; Thompson et al. 2008). The course of the disease is highly variable, ranging from asymptomatic infection to acute or chronic diarrheic illness (Liu et al. 2012). The laboratory diagnosis of *Giardia* spp. still relies mainly on a microscopy-based demonstration of cysts or trophozoites in stool samples, but several immunologically-based assays and molecular methods are also available for diagnosis (Verweij and Stensvold 2014). The prevalence of *Giardia* infection is markedly higher in developing regions of the world where *Giardia* is common in both children and adults. In recognition of the burden of disease caused by *Giardia* and to underline its link to poverty, the WHO has included it in the list of neglected disease since 2004. *G. duodenalis* has been detected in humans and animals in all regions of Southeast Asia. The distribution of *G. duodenalis* infection is shown in Table 3.1.

The pathogenesis of giardiasis is only partially known. *Giardia* is able to attach to the surface of the intestinal villi. It is assumed that a severe *Giardia* infection involves a layer of trophozoites which covers many regions of the gut wall and thus interferes with the absorption of nutrients. The mechanisms by which *G. duodenalis* produces chronic diarrhea and malabsorption remain to be clearly defined. Many infections are associated with mild to moderate mucosal damage as evidenced by animal models of infection. Possible mechanisms include direct physical injury, release of parasite products such as proteinases or lectins, and mucosal inflammation associated with T cell activation and cytokine release. Other possible mechanisms of malabsorption include associated bacterial overgrowth and bile salt deconjugation, bile salt uptake by the parasite with depletion of intraluminal bile salts, and inhibition of pancreatic hydrolytic enzymes. Thus, there is no single mechanism to explain the diarrhea and malabsorption associated with giardiasis, and currently, the pathogenesis should be regarded a multifactorial process (Farthing 1993).

Although progress has been made with regard to the diagnosis of giardiasis using non-morphology-based methods, examination (usually of feces) by microscopy generally remains the backbone of the diagnosis of intestinal parasitic infections overall owing to the fact that this method enables the detection of large number of parasitic species, including helminths. In Europe, standard methods in the laboratory diagnostics of enteric parasitoses, including *Giardia*, comprise direct microscopic examination of wet mounts and examination of concentrated feces. There is, however, an increasing trend toward the routine use of fixatives, permanent staining, and multiple sampling techniques, as these methods have been proven to considerably enhance the diagnostic yield. Advances in the biomedical sciences, however, have resulted in morphology-independent laboratory diagnosis of giardiasis. Immunoassays for the detection of copro-antigens are gradually being introduced in routine diagnostic laboratories, particularly so in laboratories processing large numbers of samples. PCR-based methods not only have superior sensitivity and specificity, but

**Table 3.1** The distribution of *G. duodenalis* infections in humans in different regions of Southeast Asia

| Country | Population/ source | Prevalence (% ) | Sample size (*n*) | References |
|---|---|---|---|---|
| *Cambodia* | | | | |
| Battambang Province, northwestern Cambodia | Schoolchildren | 31.50 | 308 | Liao et al. (2017) |
| Angkor Hospital for Children (AHC) in Siem Reap, North-Western Cambodia | Children aged under 16 years | 27.70 | 498 | Moore et al. (2016) |
| Dong village, Rovieng district, Preah Vihear Province | Humans | 18.30 | 218 | Inpankaew et al. (2014) |
| Provincial Hospital in Siem Reap | Cambodian children | 8.00 | 16,372 | Moore et al. (2012) |
| Tonle Sap Lake | School-aged children | 4.20 | 1616 | Chhakda et al. (2006) |
| Southeast of Phnom Penh | Kindergarten and Schoolchildren | 2.90 | 623 | Park et al. (2004) |
| Kampong Cham Province | Primary school Children | 3.20 | 251 | Lee et al. (2002) |
| *Indonesia* | | | | |
| Mlati, Sleman, DIY | Children and adults | 5.02 | 179 | Sari et al. (2020) |
| Dr. Soetomo General Hospital, Surabaya | HIV/AIDS patients | 5.73 | 122 | Prasetyo (2010) |
| Rancabali tea plantation, Bandung District | Children | 29.00 | 92 | Widajanti et al. (2003) |
| Jakarta | Women workers | 22.03 | 903 | Suriptiastuti (2006) |
| *Malaysia* | | | | |
| Orang Asli Selangor | Aged 2–15 years | 24.90 | 281 | Al-Mekhlafi et al. (2005) |
| Seven states of Malaysia | Children and adults | 11.6 | 1330 | Choy et al. (2014) |
| Aboriginal participants residing in Jelebu, Gerik and Temerloh States | Aged 2–74 years | 16.03 | 611 | Anuar et al. (2015) |
| Hulu Terengganu and Kemaman districts of Terengganu | Aged $\leq$ 15 years | 8.60 | 340 | Elyana et al. (2016) |
| Temerloh, Pahang | Children and adults | 12.10 and 8.29 | 473 (wet seasons = 256 and dry | Noradilah et al. (2019) |

(continued)

**Table 3.1** (continued)

| Country | Population/ source | Prevalence (% ) | Sample size ($n$) | References |
|---|---|---|---|---|
| | | | seasons = 217) | |
| Malaysia | Migrant workers in Malaysia | 10.80 | 388 | Sahimin et al. (2016) |
| *Lao PDR* | | | | |
| Rural Laos | Children and adults | 68.90 | 891 | Chard et al. (2019) |
| Lak Sip village, Luang Prabang Province | Children and adults | 6.56 | 305 | Ribas et al. (2017) |
| Mahosot Hospital, Vientiane | Children | 0.50 | 191 | Phetsouvanh et al. (1999) |
| *Thailand* | | | | |
| Pathum Thani Province | Aged 10–82 months | 37.7 | 106 | Saksirisampant et al. (2003) |
| Mae Chame district, Chiang Mai Province | Aged 3–19 years | 2.21 | 781 | Saksirisampant et al. (2004) |
| Ang Thong, Ayut-thaya, and Suphanburi Provinces | Aged 3–12 years | 1.25 | 1037 | Saksirisampant et al. (2006) |
| Sangkhlaburi, a rural district in the west of Thailand | Pre-school chil-dren 3–60 months | 23.3 | 472 | Wongstitwilairoong et al. (2007) |
| Chachoengsao Province | Primary schoolchildren | 6.20 | 531 | Ratanapo et al. (2008) |
| Villagers at the Thai/ Myanmar border | Children and adults | 2.5 | 204 | Prasertbun et al. (2012) |
| Samut Sakhon Province | Myanmar in Thailand | 14.10 | 284 | Nuchprayoon et al. (2009) |
| *Vietnam* | | | | |
| Hoa Binh | All ages | 3.00 | 2522 | Verle et al. (2003) |
| Da Nang city | 1–11 years | 20.80 | 48 | Ögren et al. (2016) |
| Da Nang city | 18–48 years | 11.30 | 80 | Ögren et al. (2016) |
| *Myanmar* | | | | |
| South Dagon and Hlaing Thar Yar dis-tricts, Yangon | Schoolchildren and guardians | 3.40 | 821 | Kim et al. (2016) |
| *The Philippines* | | | | |
| Metro Manila | Children | 11.60 | 284 | Baldo et al. (2004) |
| The Philippines | Patients | 2.00 | 3456 | Natividad et al. (2008) |
| Barangays villages | Children and adults | 19.20 | 412 | Weerakoon et al. (2018) |

also enable *Giardia* assemblage identification; however, only a few diagnostic laboratories are equipped well enough to undertake these methods. These new methods may aid diagnosis and should be considered for adoption into the routine diagnostic algorithm. Once they have been evaluated fully and compared with accepted methods in individual laboratories or through inter-laboratory quality assurance investigations, they can be regarded as an adjunct to the conventional detection and identification techniques used for laboratory diagnosis (Monis et al. 1999, 2009)

Fenbendazole and metronidazole are the most common drugs used to treat *Giardia* infections. Many medicinal products combine several active substances and often include febantel, which is converted into fenbendazole by metabolism. Thus, products containing febantel are efficacious, but only if the treatment lasts long enough. A clinical cure of diarrhea is desired. However, some infections are self-limiting, and it is reasonable to forgo treatment for patients with mild symptoms and/or contraindications to antimicrobial therapy. Metronidazole has been the drug of choice to treat giardiasis at a single high dose or multiple low doses. Tinidazole is an alternative drug used as single dose or as multiple doses. Albendazole, the drug of choice for several types of helminthiasis, has been found to be effective for the treatment of giardiasis (Gardner and Hill 2001). Lalle and Hanevik (2018) reported that clinical giardiasis is heterogeneous, with a high variability in the severity of clinical disease. It can become chronic or may be followed by post-infectious sequelae. An alarming increase in cases refractory to the conventional treatment with nitroimidazoles (i.e., metronidazole) has been reported in low prevalence settings, such as in European Union countries and especially in patients returning from Asia.

## 3.3 *Entamoeba histolytica*

*Entamoeba histolytica* is a non-flagellated protozoan enteropathogen infecting around 50 million individuals globally (World Health Organization (WHO) 1997a, b). It can cause both intestinal and extraintestinal amoebiasis. Infection arises primarily from fecal-oral transmission through the consumption of contaminated drinking water or food containing cysts. Upon ingestion, the cysts multiply into trophozoites and colonize the colonic mucosa of the host, using lectin and cysteine proteases as virulence factors, leading to host invasion (Kantor et al. 2018). The size of the trophozoite (Fig. 3.3a) is usually 15–20 μm. It may contain erythrocytes that apparently have been ingested when coming into contact to erythrocyte, which, however, are not common in the lumen of the intestine. The diameter of the cysts (Fig. 3.3b), which are released into the environment by fecal shedding, is 10–14 μm. A fully developed cyst contains four nuclei. Cigar-shaped chromatoid bodies may be seen inside the cyst too (Ryan and Ray 2004).

Infection is acquired upon the ingestion of cysts from fecally contaminated food or water. Because of the protection conferred by their walls, cysts may survive for

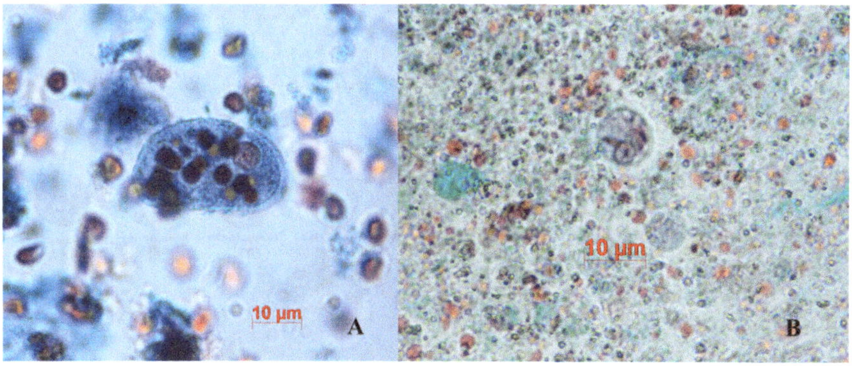


**Fig. 3.3** *Entamoeba histolytica* trophozoite with ingested red blood cells (**a**) and an immature cyst (**b**) stained with trichrome (Photo by Kotchaphon Vaisusuk)

days to weeks in the external environment. When a cyst of *E. histolytica* reaches the lower part of the ileum or cecum, excystation occurs, and an amoeba with four nuclei emerges. This divides by binary fission to form eight trophozoites. These migrate to the large intestine and may lodge in lumen or submucosal tissues. These trophozoites grow and multiply by binary fission in the large intestine. A number of trophozoites are discharged into the lumen of the bowel and are transformed into cystic forms, which are unable to excyst in the same host and therefore rely on transfer to another susceptible host via a phase in the external environment (Fig. 3.4).

Infection by *E. histolytica* occurs by ingestion of mature four-nucleated cysts in fecally contaminated food or drinking water. Excystation occurs in the small intestine and trophozoites are released, which migrate to the large intestine. The trophozoites multiply by binary fission and produce cysts, and both stages may be passed in the feces. Cysts may survive for days to weeks in the external environment and are responsible for further preparation. Trophozoites passed in the stool are rapidly destroyed once outside the body. If ingested they would not survive exposure to the gastric environment. Hence, the ingestion of trophozoites alone could not lead to an infection. In many cases, some trophozoites remain confined to the intestinal lumen of individuals which may act as asymptomatic carriers, passing only a small number of cysts in their stool. However, in a minor proportion of the cases, trophozoites invade the intestinal mucosa resulting in invasive disease and commensurate pathological manifestations.

*Entamoeba histolytica* predominantly infects people and other primates. This parasite is worldwide responsible for up to 100,000 deaths per year in the human population (World Health Organization Initiative for Vaccine Research: Parasitic Diseases 2011), thus placing it second to malaria in mortality due to protozoan parasites. Previously, it was thought that 10% of the world population was infected with *E. histolytica*, but after the detection of the species *E. dispar*, at least 90% of these infections are thought to represent infections by this non-pathogenic species

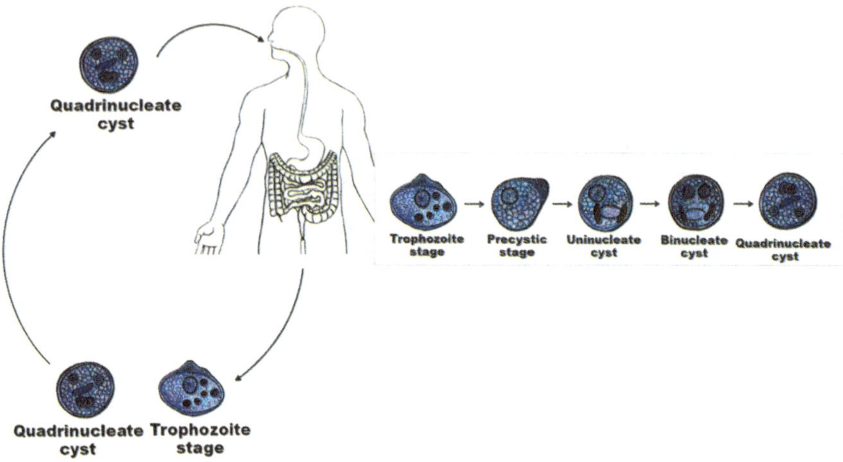


**Fig. 3.4** Cysts and trophozoites of *Entamoeba histolytica* are passed in fecal material. Cysts are typically found in formed stool, whereas trophozoites are typically found in diarrheal stool (Picture by Naruemon Bunchom)

(World Health Organization (WHO) 1997a, b). The distribution of *E. histolytica* infection in humans in different regions of Southeast Asia is shown in Table 3.2.

*Entamoeba histolytica* infection can be divided into symptomatic and asymptomatic cases. Patients may present with intestinal amoebiasis, including amoebic colitis with diarrhea, abdominal pain, and tenderness. Sometimes the signs resolve spontaneously without intervention, and sometimes the infection becomes chronic. The clinical signs of amoebiasis vary. The onset is often gradual, with patients reporting several weeks of symptoms. Profuse, watery diarrhea might be noted. *E. histolytica* may invade the colonic mucosa, and even if no blood is seen, occult blood in stools is almost always positive (Samuel and Stanley 2003). Generally, simultaneous colonic infection is seen in 50% of patients, presenting with ulcers commonly near the ileocecal valve and cecum (Bhatia and Sundaram 2019). Patients may also present with nausea, vomiting, weakness, weight loss, and referred pain to the shoulder in some cases. Patients may or may not present with jaundice (Bansal et al. 2016).

Diagnosis of *E. histolytica* relies in part on the detection of trophozoites or cysts in stool or colonic mucosa of patients. Direct smear examination on stained materials is performed microscopically. The presence of hematophagous amoebic trophozoites and cysts in a stool sample suggests an *E. histolytica* infection (González-Ruiz et al. 1994). As *Entamoeba* trophozoites generally degenerate rapidly in unfixed stool samples and refrigeration is not recommended, samples should be preserved with in a fixative (sodium acetate–acetic acid–formalin (SAF), 10% formalin, or modified polyvinyl-alcohol (PVA) which preserves the parasite's morphology and allows concentration and permanent smears to be performed. Stool culture followed by isoenzyme analysis has been considered as the "gold standard" for many years (Strachan et al. 1988). However, PCR methods become more popular and have been increasingly applied due to the possibility to differentiate *E. histolytica* from

**Table 3.2** The distribution of *E. histolytica* infection in humans in different regions of Southeast Asia

| Country | Population/ source | Prevalence (%) | Sample size (*n*) | References |
|---|---|---|---|---|
| *Cambodia* | | | | |
| Battambang Province | Schoolchildren | 17.50 | 308 | Liao et al. (2017) |
| Preah Vihear Province | Children and adults | 31.60 | 218 | Schär et al. (2014) |
| Kampong Cham Province | Primary school children | 0.80 | 251 | Park et al. (2004) |
| *Indonesia* | | | | |
| Dr. Soetomo General Hospital, Surabaya | HIV/AIDS patients | 61.50 | 122 | Prasetyo (2010) |
| Jakarta | Women workers | 14.53 | 903 | Suriptiastuti (2006) |
| *Malaysia* | | | | |
| Malaysia | Migrant workers in Malaysia | 11.60 | 388 | Sahimin et al. (2016) |
| Orang asli, Malaysia | Children and adults | 3.20 | 500 | Anuar et al. (2012) |
| Peninsular Malaysia | Children and adults | 9.15 | 426 | Ngui et al. (2012) |
| Orang asli, Malaysia | Children | 22.50 | 71 | Hartini and Kamel (2009) |
| Sabah | Children and adults | 21.00 | 150 | Aza et al. (2003) |
| *Lao PDR* | | | | |
| Lak Sip village, Luang Prabang Province | Children and adults | 2.62 | 305 | Ribas et al. (2017) |
| *Thailand* | | | | |
| Saraburi Province | 4–15 years | 1.10 | 263 | Assavapongpaiboon et al. (2018) |
| Mueang district, Nakhon Ratchasima Province | Children and adults | 11.70 | 214 | Kitvatanachai et al. (2008) |
| Institutions for mentally handicapped people | Children and adults | 7.10 | 993 | Sirivichayakul et al. (2003) |
| *Vietnam* | | | | |
| Central Vietnam | Adults | 11.20 | 383 | Blessmann et al. (2003) |
| *Myanmar* | | | | |
| South Dagon and Hlaing Thar Yar districts, Yangon | Schoolchildren and guardians | 1.20 | 821 | Kim et al. (2016) |
| Laiza, Myanmar | Adults | 15.17 | 903 | Sun et al. (2012) |
| *The Philippines* | | | | |
| Barangays (villages), the Philippines | Children and adults | 12.10 | 412 | Weerakoon et al. (2018) |
| Metro Manila, the Philippines | Children | 2.90 | 284 | Baldo et al. (2004) |

non-pathogenic amoeba. Diagnosis can additionally be reached by endoscopy of the large intestine and analysis of tissue biopsies taken. Some 89.5–100% of patients with *E. histolytica* infection are seropositive, and most patients with an amoebic liver abscess have leukocytosis, raising creative protein, positive serology, and negative blood cultures (Ferreira et al. 2020).

The WHO guidelines recommend that drinking water should be prevented from becoming contaminated by human waste and that adequate protection is provided during the transportation of potable water supplies. The boiling of drinking water for 1 min or the addition of iodine to drinking water supplies is highly recommended to sanitize water sources from *E. histolytica* cysts prior to human consumption (World Health Organization (WHO) 2019). Health education in inculcating healthy personal habits, sanitary disposal of feces, and hand washing are important control measures. Although waterborne transmission of *Entamoeba* is lower than that of other intestinal protozoans, protecting water supplies from being contaminated will lower considerably endemicity and epidemics. The treatment of amoebiasis may typically include the administration of nitroimidazole derivatives (metronidazole, tinidazole, ornidazole). Human amoebic colitis is treated with metronidazole (both luminal and tissue amoebicides), followed by a luminal agent (paromomycin, iodoquinol or diloxanide furoate) to eradicate colonization (Samuel and Stanley 2003).

## 3.4 *Cryptosporidium*

*Cryptosporidium* is a tiny protozoan parasite that infects the microvillous region of epithelial cells in the digestive and respiratory tract of vertebrates. It is also an obligate intracellular parasite of humans and other mammals, birds, reptiles, and fish. Environmentally resistant oocysts shed by infected hosts can survive adverse conditions in the environment for months until ingested by a new suitable host. Oocysts of *C. hominis/C. parvum* are spherical with a diameter of 4–6 μm. Thick- and thin-walled oocysts are formed (Fig. 3.5). The thin-walled oocysts may excyst within the same host and thus start a new life cycle (autoinfection). This may lead to a heavily infected epithelium of the small intestine, resulting in increasing malabsorptive effects and secretory diarrheas. The thick-walled oocyst is excreted with the feces and is environmentally robust.

People are infected orally by *Cryptosporidium* oocysts found in their own or other people's feces. This is the environmentally resistant transmission life cycle stage of the parasite, due to a very robust oocyst wall that protects the four sporozoites against physical and chemical damages. The infectious dose is very low and the prepatent period varies from days to weeks. Sporozoites being released inside the intestine from the oocyst penetrate the gut epithelial cells and multiply through asexual and sexual cycles, resulting in oocysts that have already sporulated inside the host and are readily infectious when they are set free in the environment (Fig. 3.6).

*Cryptosporidium*, a waterborne parasite belonging to the subphylum Apicomplexa, has been recognized as a major cause of an increasing number of

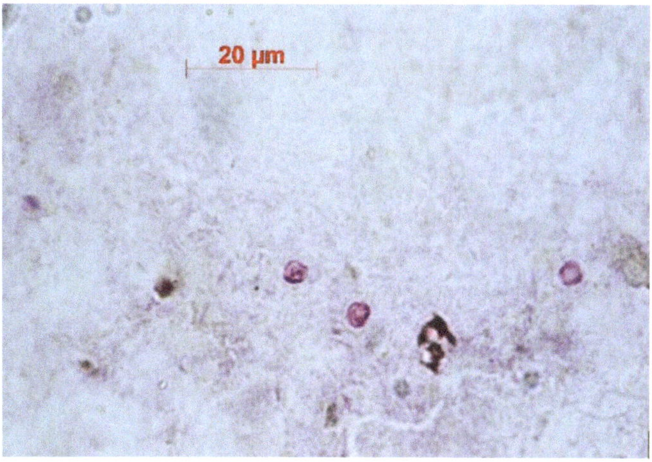


**Fig. 3.5** *Cryptosporidium* spp. oocysts in a fecal sample are stained red with a modified acid-fast stain (Photo by Kotchaphon Vaisusuk)

water-associated disease outbreaks in the past 10 years. It is transmitted through the intake of food and water contaminated with transmissible oocysts (Slifko et al. 2000; Karanis et al. 2007; Efstratiou et al. 2017). Cryptosporidiosis has its highest health impact in developed nations through foodborne, waterborne, or direct person-to-person transmission, as well as by zoonotic outbreaks of diarrheas (Yoder and Beach 2010). In developing countries, cryptosporidiosis is also one of the main causes of diarrhea and malnutrition in children and AIDS patients (Xiao and Ryan 2015). To date, isolates of *Cryptosporidium* have been assigned to at least 27 species and more than 60 genotypes that cannot be distinguished based just on morphology (Fayer and Santin 2009; Fayer 2010; Traversa 2010; Elwin et al. 2012; Ren et al. 2012; Kváč et al. 2013). *Cryptosporidium* spp. live in the brush borders of the gastrointestinal, respiratory, and renal epithelium of different vertebrates where they introduce enterocolitis, diarrhea, and cholangiopathy in humans (Chalmers and Davies 2010). *Cryptosporidium parvum* is a major cause of diarrheal excretions of humans worldwide. Global statistics on the prevalence of *C. parvum* shows that it infects up to 50 million people worldwide (World Health Organization Initiative for Vaccine Research: Parasitic Diseases 2011) depending on the local hygienic situation. In Asia and Africa, the infection occurs in from 5 to 10 million people. The epidemiology of human cryptosporidiosis and its associated species of *Cryptosporidium* can be very complex. *Cryptosporidium hominis* is hypothesized to be specific for humans and thus might be transmitted exclusively via anthroponotic pathways. In contrast, *C. parvum* appears to be capable of exploiting anthroponotic or zoonotic transmission routes, with infected cattle or small ruminants (sheep or goats) acting as reservoir hosts (Caccio 2005; Thompson et al. 2008; Robertson 2009). Improved means of transport, increased inter- and intra-country migrations and a burgeoning tourism trade have also enhanced the potential for the spread of some infectious

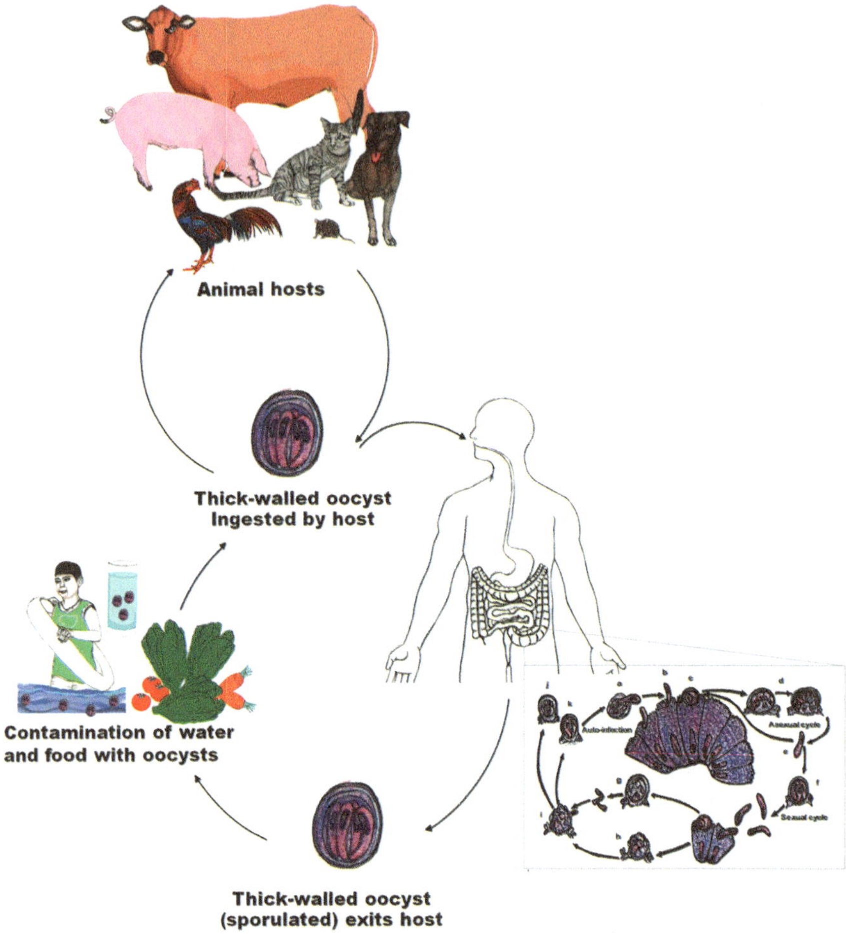


**Fig. 3.6** Sporulated oocysts, containing four sporozoites, are excreted by infected hosts within feces and possibly also via other routes such as respiratory secretions. Transmission of *Cryptosporidium* cysts occurs mainly by contact with contaminated water (e.g., drinking or recreational water) or food. Many outbreaks have probably showed in waterparks, community swimming pools, as day care centers. Zoonotic and anthroponotic transmissions of *Cryptosporidium* occur by contract to infected animals or exposure to water contaminated by the feces of infected animals. Following ingestion (and possibly inhalation) by a suitable host, excystation occurs inside its intestine (Photo by Naruemon Bunchom)

diseases (particularly those with direct modes of transmission), including cryptosporidiosis. Presently, a number of Southeast Asian countries face a severe, and likely underestimated, problem with HIV/AIDS, due to its high prevalence and rapid spread, with patients being susceptible to severe cryptosporidiosis (Anonymous 2008). The distribution of *Cryptosporidium* infection is shown in Table 3.3.

**Table 3.3** The distribution of *Cryptosporidium* infection in humans in different regions of Southeast Asia

| Country | Population/ source | Prevalence (%) | Sample size (*n*) | References |
|---|---|---|---|---|
| *Cambodia* | | | | |
| Angkor Hospital for Children (AHC) in Siem Reap, North-Western Cambodia | Children | 7.70 | 498 | Moore et al. (2016) |
| *Indonesia* | | | | |
| Jakarta, Indonesia | HIV/AIDS patients | 4.90 | 318 | Kurniawan et al. (2009) |
| General Hospital, Indonesia | HIV/AIDS patients | 52.50 | 122 | Prasetyo (2010) |
| *Malaysia* | | | | |
| Malaysia | Migrant workers | 3.10 | 388 | Sahimin et al. (2016) |
| Hospitals, Malaysia | HIV/AIDS patients | 5.20 | 346 | Iqbal et al. (2012) |
| *Lao PDR* | | | | |
| Hospitals, Lao PDR | HIV/AIDS patients | 6.60 | 137 | Paboriboune et al. (2014) |
| *Thailand* | | | | |
| Rural areas in Thailand | Children and adults | 0.14 | 697 | Prasertbun et al. (2019) |
| Hospitals in Khon Kaen Province | HIV-sero-positive patients | 11.5 | 78 | Pinlaor et al. (2005) |
| Bamrasnaradura Hospital | HIV-infected patients | 12.8 | 156 | Saksirisampant et al. (2002) |
| *The Philippines* | | | | |
| Barangays (villages), the Philippines | Children and adults | 21.80 | 412 | Weerakoon et al. (2018) |

Cryptosporidiosis in immunocompetent patients is either asymptomatic or causes self-limiting diarrhea, which may be accompanied with abdominal cramps and mild fever. In the acute phase, the diarrhea is watery. In children under 2 years patients may show severe dehydration and increased diarrhea (Flanigan et al. 1992). Cryptosporidiosis can cause outbreaks of travelers' diarrhea. The incubation period for symptoms of illness is approximately 7 days (range 1–14 days), and the illness is usually self-limiting, with a mean duration of 6–9 days. Relapses are common; reports indicate 1–5 additional episodes in 40–70% of patients. The predominant symptom is diarrhea, watery, sometimes with mucus but rarely bloody, and sometimes profuse. The pathogenesis is not well understood, but transport affecting the intestinal epithelium is probably the main cause of the diarrhea. Other symptoms include nausea, abdominal cramps, vomiting, fatigue, loss of appetite, and fever. The shedding of oocysts may continue after the cessation of the disease symptoms.

Immunocompetent individuals clear the infection and *Cryptosporidium*-associated mortality is rare among the immunocompetent population. The studies on *Cryptosporidium* in humans show a clear distinction between infection and illness. Infection and colonization of the pathogens within the intestinal tract can occur without any symptoms of gastro-enteritis (Percival et al. 2004). Protracted infection lasting for months or years may spread to the hepatobiliary tract, causing cholangiohepatitis, cholecystitis or choledochitis, or to the pancreatic duct, leading to pancreatitis (Hayward et al. 1997).

The diagnosis of *Cryptosporidium* infection is based on oocysts in fecal samples which are colored by modified acid-fast stain (AFS), which is the most common staining technique used (Weber et al. 1991). Several commercial ELISA or immunofluorescence assays have been developed for *Cryptosporidium* diagnosis (Garcia et al. 2003). In addition, PCR technologies provide a specific diagnosis up to species and genotype levels with high sensitivity (Okhuysen et al. 2001). Specific medical therapy has been developed for cryptosporidiosis. Antiparasitic drugs including nitazoxanide, paromomycin, macrolide, spiramycin, azithromycin, and rifaximin have been approved to treat cryptosporidiosis; however, their efficacy is still limited, specifically in severely affected immunocompromised people, and drug treatment is uncertain with probably limited efficacy. Infection responds best to an improved host immune status, for example, by means of HAART (Rossignol et al. 1998; Huang and White 2006).

## 3.5 *Blastocystis* sp.

*Blastocystis* is a single-celled parasite colonizing the intestinal tract of humans and numerous animal hosts, including insects and reptiles. Transmission occurs through the fecal-oral pathway, mainly by ingesting food or drink contaminated with *Blastocystis* cysts and potentially by close contact with human and non-human hosts colonized by *Blastocystis*. *Blastocystis* exists in various morphological forms. These include, but may not be limited to, the vacuolar, granular, amoeboid, and cyst forms. The size of the vacuolar form ranges between 2 and 200 μm in diameter, and the number of nuclei also varies too, with an average of four nuclei per cell (Fig. 3.7). The cyst form is generally smaller than the vacuolar form, with sizes ranging between 2 and 5 μm in diameter. Similar to the vacuolar form, a thin surface coat is also observed in the cyst form. It may survive for up to 19 days in water at normal temperature, up to one month at 25 °C, and for as long as 2 months at 4 °C. The granular form is somewhat similar to the vacuolar form, but is distinguished by the presence of multiple granules, especially within the central vacuole. Their diameters range between 3 and 80 μm. The granules might appear as myelin-like inclusions, droplets of lipid, tiny vesicles, or crystalline granules. The size of the amoeboid form is considerably smaller, measuring between 2 and 7 μm. Studies suggest the presence of extended pseudopodia, a Golgi apparatus, and mitochondria-like structures within the cytoplasmic extensions of the pseudopods. Unlike ameba,

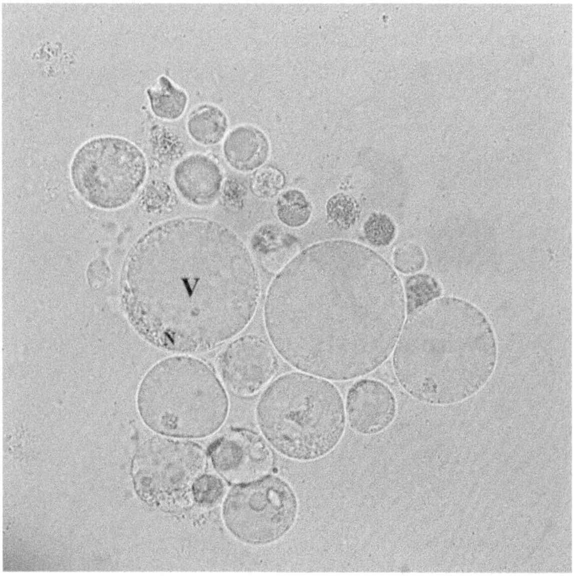


**Fig. 3.7** Light microscope image of *Blastocystis* sp. shapes from fecal culture (40× magnification). *V* vacuole, *N* nucleus (Photo by Kotchaphon Vaisusuk)

the pseudopods in *Blastocystis* are not responsible for locomotion (Dunn et al. 1989; Tan et al. 2001; Yoshikawa et al. 2004; Tan 2008).

After ingestion of the cyst, the parasite undergoes excystation in the terminal ileum, cecum, and/or large intestine of humans and develops into vacuolar forms. Encystation occurs during the passage along the large intestines, and cysts are then voided in the feces. The fecal cysts (Fig. 3.8) may be covered by a fibrillar layer, which is gradually lost during cyst development (Moe et al. 1997; Zaman et al. 1997).

The most common shape found in human stool is the cyst, which varies tremendously in size reaching from 6 to 40 μm. The thick-walled cyst shape is present in the stools is believed to be responsible for external transmission on the fecal-oral route by ingestion of contaminated water or food, exposure to infected animals or exposure to water contaminated by feces of infected animals. The cysts infect epithelial cells of the digestive tract and multiply asexually. Vacuolar forms of the parasite lead to multivacuolar and amoeboid shapes. The multivacuolar shape develops into a precyst that subsequently develops into a thin-walled cyst, which is thought to be responsible for autoinfection. The amoeboid form develops into a precyst, which later develops into a thick-walled cyst by schizogony. The thick-walled cyst is excreted in the feces.

*Blastocystis* is considered as a cosmopolitan enteric protist with a worldwide distribution. It is the most frequently detected during micro-eukaryote epidemiological surveys. Many carriers probably suffer with no or little abdominal or intestinal discomfort. Some authors report that symptoms caused by this protist comprise abdominal pain, constipation, diarrhea, flatulence, and irritable bowel syndromes (IBS). Prevalence varies widely from country to country and within various

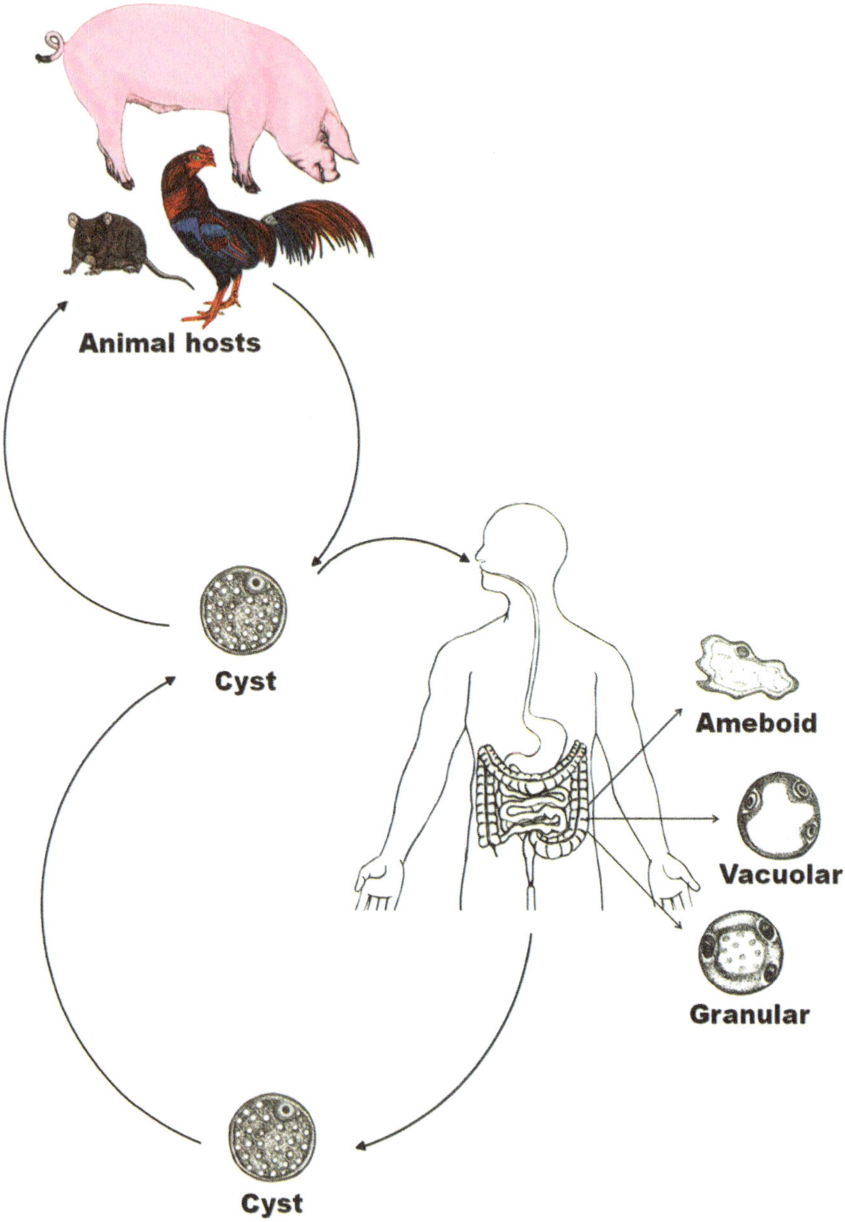


**Fig. 3.8** The life cycle and transmission of *Blastocystis hominis* (Photo by Naruemon Bunchom)

communities of the same country. In general, developing countries have a higher prevalence of the parasite than developed countries. This has been linked to poor hygiene, exposure to animal feces, and the consumption of contaminated food or water. *Blastocystis* infections are ubiquitous in people as well as in animals. Many host species have their specific *Blastocystis* species suggesting phases of co-evolution. The prevalence reported varies according to geographical region, but is generally higher in developing countries (22.1–100%) than in developed countries (0.5–23.1%), which might be related to differences in hygiene standards, type of waste disposal, frequency of exposure to animals, and the consumption of contaminated uncooked food or water (Ramírez et al. 2014; Seyer et al. 2017). Based on SSU-rDNA genotyping, high genetic variability was observed for *Blastocystis* spp., and 17 known subtypes (ST1-ST17) have been reported. Of these, subtypes 1–9 have been found in humans, with ST1–ST4 as the prevalent subtypes identified in >90 of investigations. Some human subtypes were also observed in animals, e.g., ST3 in non-human primates, ST5 in cattle and pigs, and ST7 in birds. Additionally, ST5 was commonly detected in pigs and their in-contact handlers (piggery staff) in Australia, indicating the zoonotic potential of this subtype (Li et al. 2007; Moosavi et al. 2012; Ronald et al. 2012; Alfellani et al. 2013; Roberts et al. 2013; Mattiucci et al. 2016). The high prevalence of *Blastocystis* and its pathogenic potential are indicative of its importance in Southeast Asia. The distribution of *Blastocystis* infection is shown in Table 3.4.

Symptoms commonly attributed to infection with *Blastocystis* are nonspecific and include diarrhea, abdominal pain, cramps, discomfort, and nausea. Profuse watery diarrhea has been reported in acute cases, although this may be less pronounced in chronic cases. Fatigue, anorexia, flatulence, and other nonspecific gastrointestinal effects may also be associated with *Blastocystis* infections. Fever has been reported, particularly in acute cases, but has not been noted in other studies. Other reported signs and symptoms sometimes include occurence of leukocytes in feaces, rectal bleeding, eosinophilia, hepatomegaly and splenomegaly, cutaneous rashes, and itching. One study has indicated that joint pains and swelling may result from infection of the synovial fluid by *Blastocystis*.

Traditionally, *Blastocystis* has been diagnosed by light microscopic examination of fecal material based on morphological features, which may either aid at or confuse the identification of the species involved. Wet mounts, either unstained or stained with iodine, may be used. Aqueous nigrosin has been used as a counterstain in wet mounts, and trichrome staining of fixed smears has been recommended for routine use in the diagnosis of *Blastocystis*. Serologic testing has been used in an attempt to identify patients with *Blastocystis* infections, however only with a very limited success. A lack of humoral immune response was found by Chen et al. (1987) using immunoblotting techniques with antigens from cultured *Blastocystis* isolated from four patients. Only immunoglobulin G was examined in the study, but further investigations may show a response with other immunoglobulin subclasses. Invasive techniques have occasionally detected *Blastocystis* in the intestine, but have not been evaluated and are not recommended as routine methods for diagnosis. Fluid aspirated during endoscopy has been used to detect *Blastocystis* in the lumen of the small intestine and in the cecum. Culture from fecal material appears to have no advantages over light microscopy of fresh fecal material for the detection of *Blastocystis*, and requires increased time, costs, and personnel. It also has been reported that

**Table 3.4** The distribution of *Blastocystis* sp. infection in humans in different regions of Southeast Asia

| Country | Population/ source | Prevalence (% ) | Sample size (n) | References |
|---|---|---|---|---|
| *Cambodia* | | | | |
| Battambang Province, Northwestern Cambodia | Schoolchildren | 4.9 | 308 | Liao et al. (2017) |
| *Indonesia* | | | | |
| Jakarta | Women workers | 6.56 | 903 | Suriptiastuti (2006) |
| Jakarta, Indonesia | HIV/AIDS patients | 72.40 | 318 | Kurniawan et al. (2009) |
| *Malaysia* | | | | |
| Perak and Pahang, Peninsular Malaysia | Children and adults | 20.40 | 500 | Anuar et al. (2013) |
| Pahang state, Malaysia | Schoolchildren | 25.70 | 300 | Abdulsalam et al. (2012) |
| Aboriginal community, Pahang, Malaysia | Children and adults | 46.2 | 473 | Noradilah et al. (2017) |
| *Lao PDR* | | | | |
| Lak Sip village, Luang Prabang Province | Children and adults | 14.80 | 305 | Ribas et al. (2017) |
| Champasak, Lao PDR | Children and adults | 51.70 | 60 | Sanpool et al. (2017) |
| *Thailand* | | | | |
| Simum subdistrict, Mueang district, Nakhon Ratchasima Province | Children and adults | 5.60 | 214 | Kitvatanachai et al. (2008) |
| *The Philippines* | | | | |
| Barangays (villages), the Philippines | Children and adults | 58.70 | 412 | Weerakoon et al. (2018) |
| Metro Manila, the Philippines | Children | 40.70 | 284 | Baldo et al. (2004) |

culture was successful only if large numbers of *Blastocystis* were present in the fecal material. Thus, cases with small numbers of organisms may not be detected. Any increase in the detection efficiency appears to be due to an increase in size and due to the "typical" vacuolar appearance of *Blastocystis* cells in culture and detection of the organisms by light microscopy. However, it must be recognized that shapes other than the vacuolar form of *Blastocystis* may be present in the human intestinal tract and found in fecal material. The detection of such forms in cultures of fecal material has not been assessed. The number of organisms present in the original sample cannot be determined investigating cultured material (Kukoschke et al. 1990; Zierdt 1991; Hazen 1993).

There is no clear evidence supporting the rationale of treating *Blastocystis* with anti-parasitic drugs. It is important to mention that currently, no drug has been identified that can consistently eradicate *Blastocystis* from the human intestinal tract. This is probably one of the main reasons why the number of successful randomized controlled treatment trials on *Blastocystis* is very limited. Antiprotozoal drugs, particularly the use of metronidazole, have been recommended for the treatment of *Blastocystis* infections. In combination with a luminal agent such as paromomycin, it might be able to ensure at least the temporary eradication of *Blastocystis* from the intestine. Thus, control measures include good personal hygiene, improvement in community sanitary facilities, and education to prevent fecal contamination of the environment and ingestion of contaminated material.

## 3.6 Conclusions

A relatively high prevalence of intestinal protozoan infections is common among children and HIV/AIDS patients in Southeast Asia, and to date these intestinal protozoa are considered of major public health importance. The difficulty experienced in accurately diagnosing intestinal protozoan infections might be a key reason why these pathogens and the diseases they cause are often neglected. The high prevalence of protozoan parasite infections in children and HIV/AIDS patients requires quick implementation of educational programs for children, their families, and their communities. Health education regarding hygienic practices at the school and community levels should be implemented to reduce the prevalence of intestinal protozoan parasites among these children. This should be done in parallel with comprehensive studies of the prevalence in geographical areas that have not been studied in Southeast Asia. As more sensitive techniques become available, for instance, the detection of parasite DNA by polymerase chain reaction and immunological antigen detection by monoclonal antibody assays, more parasitic infections will probably be detected. Currently, studies of the molecular genetic diversity of intestinal protozoa are few, therefore, in the future, more extensive molecular genetic studies should be conducted which can then form a sound basis for the development and instigation of effective treatment and prevention programs to deal with disease outbreaks.

## References

Abdulsalam AM, Ithoi I, Al-Mekhlafi HM et al (2012) Drinking water is a significant predictor of *Blastocystis* infection among rural Malaysian primary schoolchildren. Parasitology 139:1014–1020

Alfellani MA, Stensvold CR, Vidal-Lapiedra A et al (2013) Variable geographic distribution of *Blastocystis* subtypes and its potential implications. Acta Trop 126:11–18

Al-Mekhlafi HM, Azlin M, Nor Aini U et al (2005) Malnutrition and soil-transmitted helminthiases among Orang Asli children in Selangor, Malaysia. Asia Pac J Clin Nutr 14:188–194

Anonymous (2008) UNAIDS/WHO 2008 report of the global AIDS epidemic. UNAIDS/WHO, p 362

Anuar TS, Al-Mekhlafi HM, Ghani MKA (2012) Molecular epidemiology of amoebiasis in Malaysia: highlighting the different risk factors of *E. histolytica* and *E. dispar* infections among Orang Asli communities. Int J Parasitol 42:1165–1175

Anuar TS, Ghani MKA, Azreen SN et al (2013) *Blastocystis* infection in Malaysia: evidence of waterborne and human-to-human transmissions among the Proto-Malay, Negrito and Senoi tribes of Orang Asli. Parasite Vector 6:40

Anuar TS, Moktar N, Salleh FM, Al-Mekhlafi HM (2015) Human giardiasis in Malaysia: correlation between the presence of clinical manifestation and *Giardia intestinalis* is assemblage. Southeast Asian J Trop Med Public Health 46:835–843

Assavapongpaiboon B, Bunkasem U, Sanprasert V, Nuchprayoon S (2018) A cross-sectional study on intestinal parasitic infections in children in suburban public primary schools, Saraburi, the Central Region of Thailand. Am J Trop Med 98:763–767

Aza N, Ashley S, Albert J (2003) Parasitic infections in human communities living on the fringes of the Crocker Range Park Sabah, Malaysia. ASEAN Review of biodiversity and environmental conservation (ARBEC)

Baldo ET, Belizario VY, Leon WUD et al (2004) Infection status of intestinal parasites in children living in residential institutions in Metro Manila, the Philippines. Korean J Parasitol 42:67–70

Bansal A, Bansal AK, Bansal V, Kumar A (2016) Liver abscess: catheter drainage v/s needle aspiration. Int Surg J 2:20–25

Bhatia SJ, Sundaram S (2019) Amoebic liver abscess with synchronous colitis: lessons learnt in recent times. J Assoc Physicians India 67:11

Bingham AK, Meyer EA (1979) *Giardia* excystation can be induced in vitro in acidic solutions. Nature 277:301–302

Blessmann J, Ali IKM, Nu PAT et al (2003) Longitudinal study of intestinal *E. histolytica* infections in asymptomatic adult carriers. J Clin Microbiol 41:4745–4750

Caccio SM (2005) Molecular epidemiology of human cryptosporidiosis. Parassitologia 47:185–192

Cacciò S, Lalle M (2015) *Giardia*. In: Xiao L, Ryan U, Feng Y (eds) Biology of foodborne parasites. CRC Press, Boca Raton, pp 175–193

Chalmers RM, Davies AP (2010) Minireview: clinical cryptosporidiosis. Exp Parasitol 124:138–146

Chard AN, Baker KK, Tsai K et al (2019) Associations between soil-transmitted helminthiasis and viral, bacterial, and protozoal enteroinfections: a cross-sectional study in rural Laos. Parasite Vector 12:216

Chen J, Vaudry WL, Kowalewska K, Wenman W (1987) Lack of serum immune response to *Blastocystis hominis*. Lancet 330:1021

Chhakda T, Muth S, Socheat D, Odermatt P (2006) Intestinal parasites in schoolaged children in villages bordering Tonle Sap Lake, Cambodia. Southeast Asian J Trop Med Public Health 37:859–864

Choy SH, Al-Mekhlafi HM, Mahdy MA et al (2014) Prevalence and associated risk factors of *Giardia* infection among indigenous communities in rural Malaysia. Sci Rep 4:6909

Dunn L, Boreham P, Stenzel D (1989) Ultrastructural variation of *Blastocystis hominis* stocks in culture. Int J Parasitol 19:43–56

Efstratiou A, Ongerth JE, Karanis P (2017) Waterborne transmission of protozoan parasites: review of worldwide outbreaks - an update 2011-2016. Water Res 114:14–22

Elwin K, Hadfield SJ, Robinson G et al (2012) *Cryptosporidium viatorum* n. sp. (Apicomplexa: Cryptosporidiidae) among travellers returning to Great Britain from the Indian subcontinent, 2007-2011. Int J Parasitol 42:675–682

Elyana FN, Al-Mekhlafi HM, Ithoi I et al (2016) A tale of two communities: intestinal polyparasitism among Orang Asli and Malay communities in rural Terengganu, Malaysia. Parasite Vector 9:398

Farthing MJG (1993) Pathogenesis of giardiasis. Trans R Soc Trop Med Hyg 87:17–21

Fayer R (2010) Taxonomy and species delimitation in *Cryptosporidium*. Exp Parasitol 124:90–97

Fayer R, Santin M (2009) *Cryptosporidium xiaoi* n. sp. (Apicomplexa: Cryptosporidiidae) in sheep (*Ovis aries*). Vet Parasitol 164:192–200
Feng Y, Xiao L (2011) Zoonotic potential and molecular epidemiology of *Giardia* species and giardiasis. Clin Microbiol Rev 24:110–140
Ferreira TG, Gits-Muselli M, Dellière S et al (2020) *Entamoeba histolytica* DNA Detection in serum from patients with suspected amoebic liver abscess. J Clin Microbiol 58(10):e01153–e01120
Flanigan T, Whalen C, Turner J et al (1992) *Cryptosporidium* infection and CD4 counts. Ann Intern Med 116:840–842
Fletcher SM, Stark D, Harkness J et al (2012) Enteric protozoa in the developed world: a public health perspective. Clin Microbiol Rev 25:420–449
Garcia LS, Shimizu RY, Novak S et al (2003) Commercial assay for detection of *Giardia lamblia* and *Cryptosporidium parvum* antigens in human fecal specimens by rapid solid-phase qualitative immunochromatography. J Clin Microbiol 41:209–212
Garcia LS, Arrowood M, Kokoskin E et al (2018) Laboratory diagnosis of parasites from the gastrointestinal tract. Clin Microbiol Rev 31:e00025–e00017
Gardner TB, Hill DR (2001) Treatment of giardiasis. Clin Microbiol Rev 14:114–128
González-Ruiz A, Haque R, Aguirre A et al (1994) Value of microscopy in the diagnosis of dysentery associated with invasive *E. histolytica*. J Clin Pathol 47:236–239
Haileeyesus A, Beyene P (2009) Intestinal protozoan infections among HIV positive persons with and without Antiretroviral Treatment (ART) in selected ART centers in Adama, Afar and Dire-Dawa, Ethiopia. Ethic J Heal Dev 23:133–140
Hartini Y, Kamel AGM (2009) *E. histolytica/E. dispar* among aborigines at Pos Lenjang, Pahang. Jurnal Sains Kesihatan Malaysia 7:59–64
Hayward AR, Levy J, Facchetti F et al (1997) Cholangiopathy and tumors of the pancreas, liver, and biliary tree in boys with X-linked immunodeficiency with hyper-IgM. J Immunol 158:977–983
Hazen KC (1993) Controversial fungal and protozoan gastrointestinal infections. Curr Opin Infect Dis 6:77–82
Huang DB, White AC (2006) An updated review on *Cryptosporidium* and *Giardia*. Gastroenterol Clin N Am 35:291–314
Inpankaew T, Schär F, Odermatt P et al (2014) Low risk for transmission of zoonotic *Giardia duodenalis* from dogs to humans in rural Cambodia. Parasite Vector 7:412
Iqbal A, Lim YA, Surin J, Sim BL (2012) High diversity of *Cryptosporidium* subgenotypes identified in Malaysian HIV/AIDS individuals targeting gp60 gene. PLoS One 7:e31139
Kantor M, Abrantes A, Estevez A et al (2018) *Entamoeba histolytica*: updates in clinical manifestation, pathogenesis, and vaccine development. Can J Gastroenterol Hepatol 2018:4601420
Karanis P, Kourenti C, Smith H (2007) Waterborne transmission of protozoan parasites: a worldwide review of outbreaks and lessons learnt. J Water Health 5:1–38
Kim MJ, Jung BK, Cho J et al (2016) Prevalence of intestinal protozoans among Schoolchildren in Suburban Areas near Yangon, Myanmar. Korean J Parasitol 54:345–348
Kitvatanachai S, Boonslip S, Watanasatitarpa S (2008) Intestinal parasitic infections in Srimum suburban area of Nakhon Ratchasima Province, Thailand. Trop Biomed 25:237–242
Kucerova Z, Sokolova O, Demyanov A et al (2011) Microsporidiosis and cryptosporidiosisin HIV/AIDS patients in St. Petersburg, Russia: Serological identification of microsporidiaand *Cryptosporidium parvum* in sera samples from HIV/AIDS patients. AIDS Res Human Retroviruses 27:13–15
Kukoschke KG, Necker A, Müller HE (1990) Detection of *Blastocystis hominis* by direct microscopy and culture. Eur J Clin Microbiol Infect Dis 9:305–307
Kurniawan A, Karyadi T, Dwintasari SW et al (2009) Intestinal parasitic infections in HIV/AIDS patients presenting with diarrhoea in Jakarta, Indonesia. Trans R Soc Trop Med Hyg 103:892–898
Kváč M, Kestřánová M, Pinková M et al (2013) *Cryptosporidium scrofarum* n. sp. (Apicomplexa: Cryptosporidiidae) in domestic pigs (*Sus scrofa*). Vet Parasitol 191:218–227

Lalle M, Hanevik K (2018) Treatment-refractory giardiasis: challenges and solutions. Infect Drug Resist 11:1921–1933

Lee KJ, Bae YT, Kim DH et al (2002) Status of intestinal parasites infection among primary school children in Kampongcham, Cambodia. Korean J Parasitol 40:153–155

Li LH, Zhang XP, Lv S et al (2007) Cross-sectional surveys and subtype classification of human *Blastocystis* isolates from four epidemiological settings in China. Parasitol Res 102:83–90

Li W, Li Y, Song M et al (2015) Prevalence and genetic characteristics of *Cryptosporidium, Enterocytozoon bieneusi* and *Giardia duodenalis* in cats and dogs in Heilongjiang province, China. Vet Parasitol 208:125–134

Liao CW, Chiu KC, Chiang C et al (2017) Prevalence and risk factors for intestinal parasitic infection in schoolchildren in Battambang. Am J Trop Med Hyg 96:583–588

Liu A, Zhang X, Zhang L et al (2012) Occurrence of bovine giardiasis and endemic genetic characterization of *Giardia duodenalis* isolates in Heilongjiang Province, in the Northeast of China. Parasitol Res 111:655–661

Mattiucci S, Crisafi B, Gabrielli S, Paoletti M, Cancrini G (2016) Molecular epidemiology and genetic diversity of *Blastocystis* infection in humans in Italy. Epidemiol Infect 144:635–646

Moe KT, Singh M, Howe J et al (1997) Experimental *Blastocystis hominis* infection in laboratory mice. Parasitol Res 83:319–325

Monis PT, Andrews RH, Mayrhofer G, Ey PL (1999) Molecular systematics of the parasitic protozoan *Giardia intestinalis*. Mol Biol Evol 16:1135–1144

Monis PT, Caccio SM, Thompson RCA (2009) Variation in *Giardia*: towards a taxonomic revision of the genus. Trends Parasitol 25:93–100

Moore CE, Hor PC, Soeng S et al (2012) Changing patterns of gastrointestinal parasite infections in Cambodian children: 2006-2011. J Trop Pediatr 58:509–512

Moore CE, Elwin K, Phot N et al (2016) Molecular characterization of *Cryptosporidium* species and *Giardia duodenalis* from symptomatic Cambodian children. PLoS Negl Trop Dis 10: e0004822

Moosavi A, Haghighi A, Mojarad EN et al (2012) Genetic variability of *Blastocystis* sp. isolated from symptomatic and asymptomatic individuals in Iran. Parasitol Res 111:2311–2315

Natividad FF, Buerano CC, Lago CB et al (2008) Prevalence rates of *Giardia* and *Cryptosporidium* among diarrheic patients in the Philippines. Southeast Asian J Trop Med Public Health 39:991–999

Ngui R, Angal L, Fakhrurrazi SA et al (2012) Differentiating *E. histolytica*, *E. dispar* and *E. moshkovskii* using nested polymerase chain reaction (PCR) in rural communities in Malaysia. Parasite Vector 5:187

Noradilah SA, Moktar N, Anuar TS et al (2017) Molecular epidemiology of blastocystosis in Malaysia: does seasonal variation play an important role in determining the distribution and risk factors of *Blastocystis* subtype infections in the Aboriginal community? Parasite Vector 10:360

Noradilah SA, Moktar N, Lee IL et al (2019) Impact of seasonality on the prevalence and risk factors of *Giardia lamblia* infections among the aborigines. Original Article 12:315–320

Nuchprayoon S, Sanprasert V, Kaewzaithim S, Saksirisampant W (2009) Screening for intestinal parasitic infections among Myanmar migrant workers in Thai food industry: a high-risk transmission. J Immigr Minor Health 11:115–121

Ögren J, Nguyen SV, Nguyen MK et al (2016) Prevalence of *Dientamoeba fragilis*, *Giardia duodenalis*, *Entamoeba histolytica/dispar*, and *Cryptosporidium* spp. in Da Nang, Vietnam, detected by a multiplex real-time PCR. APMIS 124:529–533

Okhuysen PC, Robinson P, Nguyen MT et al (2001) Jejunal cytokine response in AIDS patients with chronic cryptosporidiosis and during immune reconstitution. AIDS 15:802–804

Ortega YR, Eberhard ML, Kris H (2008) Protozoan diseases: cryptosporidiosis, giardiasis and other intestinal protozoan diseases. In: International encyclopedia of public health. Academic Press, Oxford, pp 354–366

Paboriboune P, Phoumindr N, Borel E et al (2014) Intestinal parasitic infections in HIV-infected patients, Lao People's Democratic Republic. PLoS One 9:e91452

Park SK, Kim DH, Deung YK et al (2004) Status of intestinal parasite infections among children in Bat Dambang, Cambodia. Korean J Parasitol 42:201–203
Percival S, Chalmers R, Embrey M et al (2004) Microbiology of waterborne diseases. Elsevier, Amsterdam
Phetsouvanh R, Midorikawa Y, Nakamura S (1999) The seasonal variation in the microbial agents implicated in the etiology of diarrheal diseases among children in Lao People's Democratic Republic. Southeast Asian J Trop Med Public Health 30:319–323
Pinlaor S, Mootsikapun P, Pinlaor P et al (2005) Detection of opportunistic and non-opportunistic intestinal parasites and liver flukes in HIV-positive and HIV-negative subjects. Southeast Asian J Trop Med Public Health 36:841–845
Prasertbun R, Sukthana Y, Popruk S (2012) Real-time PCR: benefits for detection of mild and asymptomatic *Giardia* infections. Trop Med Health 40:31–35
Prasertbun R, Mori H, Sukthana Y et al (2019) *Enterocytozoon bieneusi* and *Cryptosporidium*: a cross-sectional study conducted throughout Thailand. BMC Infect Dis 19:808
Prasetyo RH (2010) Intestinal parasites infection in AIDS patients with chronic diarrhea at Dr. Soetomo General Hospital Surabaya. Indones J Trop Infect Dis 1:36–37
Ramírez JD, Sánchez LV, Bautista DC et al (2014) *Blastocystis* subtypes detected in humans and animals from Colombia. Infect Genet Evol 22:223–228
Ratanapo S, Mungthin M, Soontrapa S et al (2008) Multiple modes of transmission of giardiasis in primary schoolchildren of a rural community, Thailand. Am J Trop Med Hyg 78:611–615
Ren X, Zhao J, Zhang L et al (2012) *Cryptosporidium tyzzeri* n. sp. (Apicomplexa: Cryptosporidiidae) in domestic mice (*Mus musculus*). Exp Parasitol 130:274–281
Rendtorff RC (1954) The experimental transmission of human intestinal protozoan parasites. II. *Giardia lamblia* cysts given in capsules. Am J Hyg 59:209–220
Ribas A, Jollivet C, Morand S et al (2017) Intestinal parasitic infections and environmental water contamination in a rural village of northern Lao PDR. Korean J Parasitol 55:523–532
Roberts T, Stark D, Harkness J, Ellis J (2013) Subtype distribution of *Blastocystis* isolates from a variety of animals from New South Wales, Australia. Vet Parasitol 196:85–89
Robertson LJ (2009) *Giardia* and *Cryptosporidium* infections in sheep and goats: a review of the potential for transmission to humans via environmental contamination. Epidemiol Infect 137:913–921
Ronald F, Monica S, Dumitru M (2012) Detection of concurrent infection of dairy cattle with *Blastocystis, Cryptosporidium, Giardia*, and *Enterocytozoon* by molecular and microscopic methods. Parasitol Res 111:1349–1355
Rossignol JF, Hidalgo H, Feregrino M et al (1998) A double-'blind' placebo-controlled study of nitazoxanide in the treatment of cryptosporidial diarrhoea in AIDS patients in Mexico. Trans R Soc Trop Med Hyg 92:663–666
Ryan KJ, Ray CG (2004) Sherris medical microbiology, 4th edn. McGraw Hill, New York, pp 733–738
Sahimin N, Lim YA, Ariffin F et al (2016) Migrant workers in Malaysia: current implications of sociodemographic and environmental characteristics in the transmission of intestinal parasitic infections. PLoS Negl Trop Dis 10:e0005110
Saksirisampant W, Eampokalap B, Rattanasrithong M et al (2002) A prevalence of *Cryptosporidium* infections among Thai HIV infected patients. J Med Assoc Thail 85:S424–S428
Saksirisampant W, Nuchprayoon S, Wiwanitkit V (2003) Intestinal parasitic infestations among children in an orphanage in Pathum Thani province. J Med Assoc Thail 86:S263–S270
Saksirisampant W, Prownebon J, Kanmamee S et al (2004) Prevalence of parasitism among students of the Karen Hill-Tribe in Mae Chame District, ChiangMai Province, Thailand. J Med Assoc Thail 87:278–283
Saksirisampant W, Prownebon J, Kulkunthorn M et al (2006) The prevalence of intestinal parasitic infections among school children in the central region of Thailand. J Med Assoc Thail 89:1928–1933
Samuel L, Stanley JR (2003) Amoebiasis. Lancet 361:1025–1034

Sanpool O, Laymanivong S, Thanchomnang T et al (2017) Subtype identification of human *Blastocystis* spp. isolated from Lao People's Democratic Republic. Acta Trop 168:37–40

Sari Y, Suryawati B, Probandari AN et al (2020) Prevalence of giardiasis and its associated factors among livestock and rivers in Mlati, Sleman, Yogyakarta, Indonesia. Bali Med J 9:266–270

Schär F, Inpankaew T, Traub RJ et al (2014) The prevalence and diversity of intestinal parasitic infections in humans and domestic animals in a rural Cambodian village. Parasitol Int 63:597–603

Seyer A, Karasartova D, Ruh E et al (2017) Epidemiology and prevalence of *Blastocystis* spp. in North Cyprus. Am J Trop Med Hyg 96:1164–1170

Sirivichayakul C, Pojjaroenanant C, Wisetsing P (2003) Prevalence of intestinal parasitic infection among Thai people with mental handicaps. Southeast Asian J Trop Med Public Health 34:259–263

Slifko TR, Smith HV, Rose JB (2000) Emerging parasite zoonoses associated with water and food. Int J Parasitol 30:1379–1393

Strachan WD, Chiodini PL, Spice WM et al (1988) Immunological differentiation of pathogenic and non-pathogenic isolates of *E. histolytica*. Lancet 1:561–563

Sun XD, Deng Y, Wang J et al. (2012) A survey on the prevalence of amoebiasis in Western China near the China-Myanmar border. J Pathogen Biol http://en.cnki.com.cn/Article_en/CJFDTOTAL-ZISC201201020.htm

Suriptiastuti (2006) Some epidemiological aspects of intestinal parasites in women workers before going abroad. Trop Biomed 23:103–108

Tan KS (2008) New insights on classification, identification, and clinical relevance of *Blastocystis* spp. Clin Microbiol Rev 21:639–665

Tan KS, Howe J, Yap EH, Singh M (2001) Do *Blastocystis hominis* colony forms undergo programmed cell death? Parasitol Res 87:362–367

Thompson RC, Palmer CS, O'Handley R (2008) The public health and clinical significance of *Giardia* and *Cryptosporidium* in domestic animals. Vet J 177:18–25

Traversa D (2010) Evidence for a new species of *Cryptosporidium* infecting tortoises: *Cryptosporidium ducismarci*. Parasite Vector 3:21

Verle P, Kongs A, De NV et al (2003) Prevalence of intestinal parasitic infections in northern Vietnam. Tropical Med Int Health 8:961–964

Verweij JJ, Stensvold CR (2014) Molecular testing for clinical diagnosis and epidemiological investigations of intestinal parasitic infections. Clin Microbiol Rev 27:371–418

Weber R, Bryan RT, Bishop HS et al (1991) Threshold of detection of *Cryptosporidium* oocysts in human stool specimens: evidence for low sensitivity of current diagnostic methods. J Clin Microbiol 29:1323–1327

Weerakoon KG, Gordon CA, Williams GM (2018) Co-parasitism of intestinal protozoa and *Schistosoma japonicum* in a rural community in the Philippines. Infect Dis Poverty 7:121

Widajanti E, Garna H, Chairulfatah A, Hudaya D (2003) Giardiasis in daycare centers at Rancabali tea plantation, Bandung District. Paediatr Indones 43:9–10

Wongstitwilairoong B, Srijan A, Serichantalergs O et al (2007) Intestinal parasitic infections among pre-school children in Sangkhlaburi, Thailand. Am J Trop Med Hyg 76:345–350

World Health Organization (WHO) (1987) Prevention and control of intestinal parasitic infections. WHO Tech Rep 749:1–86

World Health Organization (WHO) (1997a) Amoebiasis. WHO Wkly Epidemiol Rec 72:97–100

World Health Organization (WHO) (1997b) Pan American health organization. UNESCO report of a consultation of experts on amoebiasis. Wkly Epidemiol Rec 72:97–99

World Health Organization (WHO) 2019 *Guidelines for drinking-water quality: fourth edition incorporating first addendum*, 4th ed. World Health Organization. Retrieved from https://apps.who.int/iris/handle/10665/254637. Accessed 15 Nov 2019

World Health Organization Initiative for Vaccine Research: Parasitic Diseases 2011. Available from: http://www.who.int/vaccine_research/diseases/soa_parasitic/en/index1.html. Accessed 27 April 2011

Xiao L, Ryan U (2015) *Cryptosporidium*. In: Xiao L, Ryan U, Feng Y (eds) Biology of foodborne parasites. CRC Press, New York, pp 77–95
Yoder JS, Beach MJ (2010) *Cryptosporidium* surveillance and risk factors in the United States. Exp Parasitol 124:31–39
Yoshikawa H, Yoshida K, Nakajima A et al (2004) Fecal-oral transmission of the cyst form of *Blastocystis hominis* in rats. Parasitol Res 94:391–396
Zaman V, Howe J, Ng M (1997) Variation in the cyst morphology of *Blastocystis hominis*. Parasitol Res 83:306–308
Zeibig EA (1997) Clinical parasitology: a practical approach. Sounders, New York, p 320
Zierdt CH (1991) *Blastocystis hominis*—past and future. Clin Microbiol Rev 4:61–79

# Chapter 4
# Biodiversity of Human Trematodes and Their Intermediate Hosts in Southeast Asia

**Weerachai Saijuntha, Ross H. Andrews, Paiboon Sithithaworn, and Trevor N. Petney**

**Abstract** A diverse range of human trematodes commonly dwell in the intestine, bile ducts of the liver, lung, and blood vessel infecting millions of people in continental Southeast Asia. Similarly, their life cycles involve a diverse range of intermediate hosts comprising aquatic plants, snails, fish, and other invertebrates. A sound knowledge of the biodiversity of these trematodes and their intermediate hosts is critically important for the development and establishment of future prevention and control programs on trematodiases in Southeast Asia. Therefore, this chapter will focus on the biology, life cycles, species diversity, and genetic diversity of medically important trematodes including their intermediate hosts and their distribution in Southeast Asia. The chapter will also examine the route of transmission via food and water as well as a variety of traditionally cooked dishes and raw attitudes that pose significant risks of trematode infection in people throughout Southeast Asia.

W. Saijuntha (✉)
Walai Rukhavej Botanical Research Institute, Biodiversity and Conservation Research Unit, Mahasarakham University, Maha Sarakham, Thailand
e-mail: weerachai.s@msu.ac.th

R. H. Andrews
Cholangiocarcinoma Research Institute, Faculty of Medicine, Khon Kaen University, Khon Kaen, Thailand

Faculty of Medicine, Imperial College, London, UK

P. Sithithaworn
Cholangiocarcinoma Research Institute, Faculty of Medicine, Khon Kaen University, Khon Kaen, Thailand

T. N. Petney
Departments of Zoology and Paleontology and Evolution, State Museum of Natural History, Karlsruhe, Germany

Cholangiocarcinoma Research Institute, Khon Kaen University, Khon Kaen, Thailand

 
T. N. Petney et al. (eds.), *Biodiversity of Southeast Asian Parasites and Vectors causing Human Disease*, Parasitology Research Monographs 14,
https://doi.org/10.1007/978-3-030-71161-0_4

**Keywords** Trematodes · Snail intermediate hosts · Life cycles · Transmission cycles · Species diversity

## 4.1 Introduction

Southeast Asia is one of the major biodiversity hotspots in the world. Four biodiversity hotspots have been defined and located in this region, namely the Sundaland, Wallacea, the Philippines, and Indo-Burma biodiversity hotspots (Myers et al. 2000). There is also a high diversity of trematodes and similarly of their hosts' life cycles. Southeast Asian people are susceptible to infection by at least 50 species of digenetic trematodes (those that utilize more than one host in their life cycle), including intestinal flukes, liver flukes, lung flukes, and blood flukes (Johansen et al. 2010).

Trematodes belong to the class Trematoda of the phylum Platyhelminthes that comprises two orders, monogenean and digenean trematodes. Only the digenean parasites are of medical importance to people, and these are the focus of this chapter. Almost all of them are cosmopolitan that are obligatorily parasitic infecting invertebrate intermediate and vertebrate definitive hosts. Their life cycles are complex involving at least two hosts. The first intermediate hosts are usually freshwater snails, while the second intermediate hosts are usually aquatic invertebrates and occasionally vertebrates (Toledo and Fried 2019).

In Southeast Asia, a diverse group of species of digenean trematodes have been found to infect and cause disease in people (Utzinger et al. 2010). For instance, there are liver flukes causing fascioliasis, opisthorchiasis, and clonorchiasis, lung flukes which are the causative agents of paragonimiasis, intestinal flukes causing echinostomiasis and heterophyiasis, as well as blood flukes causing schistosomiasis. Of these digenean trematodes, two species of liver flukes, *Opisthorchis viverrini* and *Clonorchis sinensis*, have been classified as group 1 carcinogens, which act as the causative agents of bile duct cancer (cholangiocarcinoma; CCA) (WHO 1995). On the other hand, paragonimiasis, which is caused by a diverse group of lung flukes in the genus *Paragonimus,* is another important foodborne trematode distributed throughout Southeast Asia. The most diverse groups of human trematodes in Southeast Asia are the intestinal flukes, echinostomatids in the family Echinostomatidae and heterophylids in the family Heterophyidae. People suffer from schistosomiasis caused by *Schistosoma japonicum* throughout the Philippines and Indonesia, whereas in some restricted areas in the lower Mekong Basin, for instance, in Lao PDR, Cambodia and in Myanmar, schistosomiasis is caused by the closely related species, *S. mekongi* (Toledo and Fried 2019).

Worldwide almost all trematodiases affect the poorest people, predominately in rural areas. Many factors contribute to the high prevalence of infections, including a lack of education, poverty, malnutrition, a lack of food inspection, and/or poor sanitation (Fried et al. 2004). As the second intermediate hosts are a highly diverse group of edible animals and the metacercariae of some species are found on food plants that are consumed as part of regular daily meals, the consumption of raw or partially cooked aquatic plants and animals is the major risk factor leading to the high levels of infection found throughout Southeast Asia (Toledo and Fried 2019).

## 4.2 Biology and Life Cycle of Trematodes

Almost all trematodes are hermaphroditic displaying obligate alternation between sexual and clonal reproduction over their life cycles, with the exception of members of the family Schisotsomatidae which have separate sexes. They have traditionally been classified based on the human organ(s) they inhabit, namely intestinal flukes, liver flukes (including pancreatic flukes), lung flukes, and blood flukes (Fig. 4.1). Structurally, trematodes are flat and elongated parasitic worms whose outer surface (tegument) contains microvilli that both protect the worm and act as a nutrient absorptive surface. Adult worms possess anterior and ventral suckers, which are useful in maintaining their attachment to host tissue. Trematodes possess a blind intestine that originates from the anterior sucker and provides additional absorptive

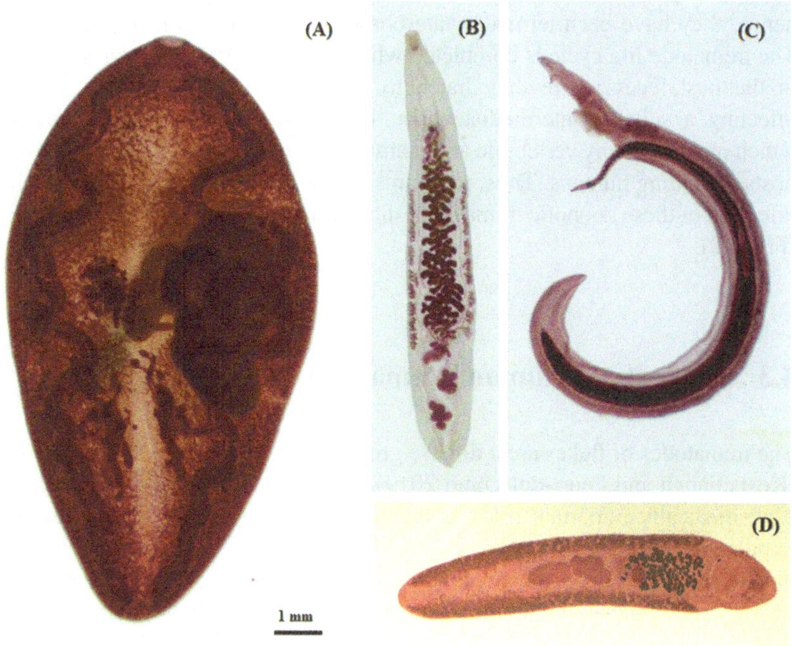


**Fig. 4.1** Adult worms of (**a**) lung fluke, *Paragonimus heterotremus*; (**b**) liver fluke, *Opisthorchis viverrini*, (**c**) blood fluke, *Schistosoma japonicum* [(**a**)–(**c**) photos courtesy by Thongchit Thanchomnang], (**d**) intestinal fluke; *Hypoderaeum conoideum* (photo by Weerachai Saijuntha)

capacity. Insoluble intestinal contents are regurgitated through the apical sucker, whereas liquid waste may be expelled through specialized excretory cells (flame cells).

The digenean life cycle of a parasitic trematode (Fig. 4.2) typically consists of the definitive host, usually a vertebrate, including people, where the flukes reproduce sexually. Although self-fertilization may occur in some species, cross-fertilization is the most extended form of reproduction during the adult stage (Carbonell and Fried 2019). The first intermediate hosts, in which asexual reproduction occurs, are typically species of aquatic snails. Asexual reproduction produces cercariae, which are motile forms of the worm. After being shed by the first intermediate hosts, cercariae infect second intermediate hosts by percutaneous penetration, or they encyst on aquatic plants until consumed by vertebrate hosts, in which further development occurs. The aquatic plants include morning glory and water mimosa, and the aquatic animal hosts, naiads, tadpoles, freshwater snails (Fig. 4.3), and crabs as well as a large number of freshwater cyprinid fish (Toledo and Fried 2019).

As almost all trematodes are transmitted to people via eating food (foodborne transmission), these trematodes are known as "foodborne trematodes." The blood flukes of the family Schistosomatidae, however, do not require a second intermediate host, but their cercariae will shed from their first snail intermediate hosts, then swim freely and penetrate directly into their primary hosts (waterborne transmission), hence, they have been termed "waterborne trematodes" (Toledo and Fried 2019). The trematode life cycle is completed when eggs shed by adult worms are excreted in the host feces where they hatch to release ciliated miracidiae, subsequently infecting a suitable intermediate host. However, most trematodes are zoonotic, which includes many vertebrate and mammal species as their definitive and reservoir hosts, including humans. Thus, programs for the effective prevention and control of infection of these zoonotic trematodes should also focus on their zoonotic life cycles (Fig. 4.2).

## 4.3 Diversity of Human Trematodes

The trematodes or flukes are a diverse group which includes 18,000–24,000 species (Kostadinova and Pérez-del-Olmo 2019). Of these, at least 54 species play significant, medically important roles as they infect large numbers of people, including populations throughout Southeast Asia (Table 4.1). Their classification has predominantly been based on the organ they infect/inhabit in people, namely intestinal flukes, liver flukes (including pancreatic fluke), lung flukes, and blood flukes (Fig. 4.1).

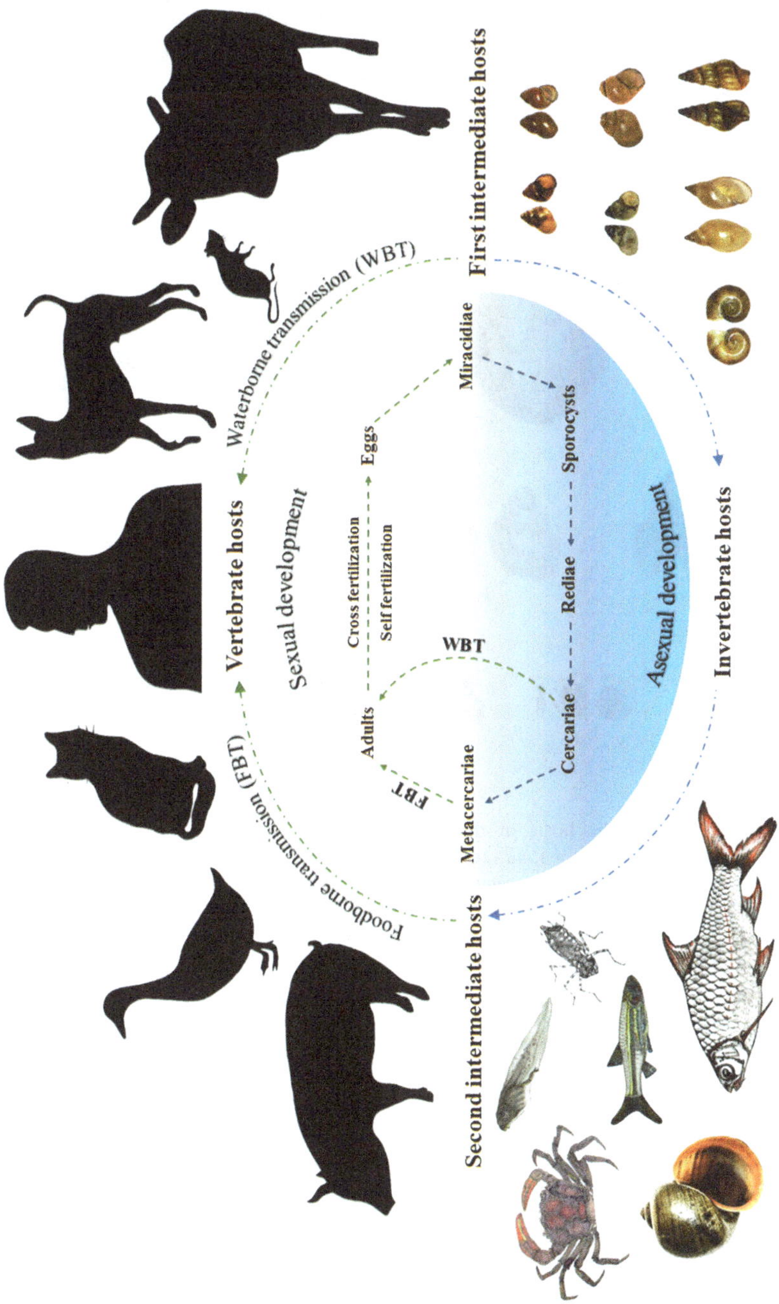


**Fig. 4.2** General life cycle of digenetic trematodes found in Southeast Asia (Illustration by Naruemon Bunchom)

**Fig. 4.3** Snail intermediate hosts of digenetic trematodes found in Thailand (**a**) *Pomacea canaliculata*; (**b**) *Pomacea maculata*; (**c**) *Pila polita*; (**d**) *Pila ampullacea*; (**e**) *Filopaludina martensi martensi*; (**f**) *Filopaludina sumatrensis polygramma*; (**g**) *Melanoides tuberculata*; (**h**) *Lymnaea (Radex) rubiginosa*; (**i**) *Indoplanorbis exustus*; (**j**) *Gyraulus convexiusculus*; (**k**) *Hydrobioides nassa*; (**l**) *Bithynia funiculata*; (**m**) *Bithynia siamensis goniomphalos*; (**n**) *Bithynia siamensis siamensis* (photo courtesy by Naruemon Bunchom)

### *4.3.1 Intestinal Flukes*

Intestinal flukes are the largest group of human trematodes, consisting of at least 41 species belonging to 22 genera and 8 families distributed throughout Southeast Asia (Chai et al. 2009a, 2018). The most diverse group are echinostomatids in the family Echinostomatidae consisting of 16 species in 7 genera, with members of the genus *Echinostoma* being the most common causative agents of human echinostomiasis in Southeast Asia (Toledo and Esteban 2016). The next diverse group comprises the heterophyids in the family Heterophyidae, which constitutes 13 species belonging to 7 genera, while the family Plagiorchiidae contains 4 species

**Table 4.1** The trematode species found in Southeast Asia together with their method of transmission and geographical distribution

| Flukes | Family | Genus | Species | Sources of human infection | Geographical distribution of human cases in Southeast Asia |
|---|---|---|---|---|---|
| Intestinal flukes | Echinostomatidae | *Artyfechinostomum* | *A. malayanum* | Freshwater snails | Indonesia, Lao PDR, Malaysia, Philippines, Singapore, Thailand |
| | | | *A. oraoni* | Freshwater snails | Thailand |
| | | | *A. sufrartyfex* | Freshwater snails | Vietnam |
| | | *Echinostoma* | *E. revolutum* | Freshwater snails | Cambodia, Lao PDR, Thailand, Vietnam |
| | | | *E. ilocanum* | Freshwater snails | Cambodia, Lao PDR, Indonesia, Malaysia, Philippines, Thailand |
| | | | *E. echinatum* | Freshwater snails | Indonesia, Lao PDR, Thailand |
| | | | *E. cinetorchis* | Freshwater snails/ fish | Vietnam |
| | | | *E. macrorchis* | Freshwater snails | Indonesia, Lao PDR |
| | | | *E. aegyptiaca* | Unknown | Lao PDR |
| | | | *E. mekongi* | Freshwater snails | Cambodia |
| | | *Echinoparyphium* | *E. recurvatum* | Freshwater snails | Indonesia, Philippines, Thailand |
| | | *Echinochasmus* | *E. japonicus* | Freshwater fish | Lao PDR, Thailand, Vietnam |
| | | | *E. perfoliatus* | Freshwater fish | Thailand, Vietnam |
| | | *Epistmium* | *E. caninum* | Freshwater fish | Thailand |
| | | *Euparyphium* | *Euparyphium* sp. | Tadpole | Lao PDR |
| | | *Hypoderaeum* | *H. conoideum* | Freshwater snails, tadpole | Indonesia, Thailand |
| | Heterophyidae | *Centrocestus* | *C. caninus* | Freshwater fish | Thailand |
| | | | *C. formosanus* | Freshwater fish | Lao PDR, Philippines, Thailand, Vietnam |
| | | *Haplorchis* | *H. pumilio* | Freshwater fish | Thailand, Vietnam |
| | | | *H. taichui* | Freshwater fish | Lao PDR, Philippines, Thailand, Vietnam |
| | | | *H. vanissimus* | Freshwater fish | Philippines |
| | | | *H. yokogawai* | Freshwater fish | Indonesia, Lao PDR, Malaysia, Philippines, Thailand |

(continued)

**Table 4.1** (continued)

| Flukes | Family | Genus | Species | Sources of human infection | Geographical distribution of human cases in Southeast Asia |
|---|---|---|---|---|---|
| | | *Heterophyopsis* | *H. continua*[a] | Blackish water fish | Philippines, Vietnam |
| | | *Procerovum* | *P. calderoni* | Freshwater fish | Philippines |
| | | | *P. varium* | Mullet | Cambodia, Lao PDR, Philippines, Vietnam |
| | | *Pygidiopsis* | *P. genata* | Mullet | Philippines |
| | | *Stellanchasmus* | *S. falcatus* | Mullet, half-beaked fish | Thailand, Philippines, Vietnam |
| | | | *S. pseudocirratus* | Mullet | Philippines |
| | | *Stritodora* | *S. lari* | Goby | Vietnam |
| | Plagiorchiidae | *Plagiorchis* | *P. harinasutai* | Probably insect larva | Thailand |
| | | | *P. muris* | Insect larva, naiad | Cambodia, Lao PDR, Philippines, Thailand, Vietnam |
| | | | *P. javensis* | Insect larva | Indonesia |
| | | | *P. philippinensis* | Insect larva | Philippines |
| | Lechithodendriidae | *Phaneropsolus* | *P. bonnei* | Freshwater fish, naiads | Cambodia, Indonesia, Lao PDR, Malaysia, Thailand |
| | | *Prosthodendrium* | *P. molenkampi* | Naiad, damselfly | Cambodia, Indonesia, Lao PDR, Thailand |
| | Isoparorchiidae | *Isoparorchis* | *I. hypselobagri* | Unknown | Indonesia, Thailand |
| | Microphallidae | *Microphallus* | *M. brevicaeca* | Shrimp | Philippines |
| | Paramphistomatidae | *Fischoederius* | *F. elongatus* | Aquatic plants | Thailand, Vietnam |
| | Fasciolidae | *Fasciolopsis* | *F. buski* | Aquatic plants | Cambodia, Indonesia, Lao PDR, Malaysia, Philippines, Singapore, Thailand, Vietnam |
| | Cathaemaciidae | *Cathaemacia* | *C. cabrerai* | Unknown | Philippines |
| | Gastrodiscidae | *Gastrodiscoides* | *G. hominis* | Tadpole, crayfish, aquatic plants | Cambodia, Indonesia, Malaysia, Myanmar, Philippines, Thailand, Vietnam |

| | | | | | |
|---|---|---|---|---|---|
| Liver flukes | Opisthorchiidae | *Opisthorchis* | *O. viverrini* | Freshwater cyprinid fish | Cambodia, Lao PDR, Myanmar, Thailand, Vietnam |
| | | *Clonorchis* | *C. sinensis* | Freshwater cyprinid fish | Vietnam |
| | Fasciolidae | *Fasciola* | *F. gigantica* | Aquatic plants | Myanmar, Thailand, Vietnam |
| | | | *F. hepatica* | Aquatic plants | Myanmar, Thailand, Vietnam |
| | Dicrocoeliidae | *Dicrocoelium* | *D. dendriticum* | Ant | Philippines |
| | | *Eurytrema* | *E. pancreaticum*[a] | Grasshopper | Malaysia |
| Lung flukes | Paragonomidae | *Paragonimus* | *P. heterotremus* | Freshwater crab | Cambodia, Lao PDR, Myanmar, Thailand, Vietnam |
| | | | *P. pseudoheterotremus* | Freshwater crab | Thailand |
| | | | *P. westermani* | Freshwater crab, crayfish | Philippines, Thailand |
| | | | *P. philippinensis* | Freshwater crab | Philippines |
| Blood flukes | Schistosomatidae | *Schistosoma* | *S. japonicum* | Water contact | Indonesia, Philippines |
| | | | *S. mekongi* | Water contact | Cambodia, Lao PDR, Myanmar |
| | | | *S. malayensis* | Water contact | Malaysia |

[a]Commonly found in pancreatic duct but rarely found in liver bile duct

in 1 genus. The family Lecithodendriidae includes two species in two genera, and the remaining six families have one species in one genus (Table 4.1) (Chai et al. 2009a). People who suffer from intestinal fluke infection can present with severe epigastric or abdominal pain with diarrhea, fatigue, malnutrition, and anemia (Toledo et al. 2019).

In mainland Southeast Asia the main disease foci are found along the Mekong River basin with a corresponding high prevalence of infection by echinostomes (Chai et al. 2020; Toledo et al. 2019). Almost all of the echinostome species have recently been reported to infect riparian people in Thailand, Lao PDR, Cambodia, and Vietnam (Chai et al. 2018). For example, *Artyfechinostomum malayanum*, *Echinostoma revolutum*, *Echinochasmus japonicus*, and *Euparyphium* sp. have recently been recovered from people in Savannakhet and Khammouane Provinces in Lao PDR (Chai et al. 2012). Most recently, *Echinostoma ilocanum* and *Echinostoma aegyptica* have been reported to infect riparian people in Savannakhet Province, Lao PDR (Chai et al. 2018, 2020). *Echinostoma mekongi* was also recently reported as a new species that is morphologically and molecularly distinct from the other members of 37 collar-spined group infecting riparian people along the Mekong River in Cambodia (Cho et al. 2020).

Heterophyids are another diverse group of intestinal flukes with the prevalence of infection in people being high in the lower Mekong River basin (Chai et al. 2009b). This is similar to *Haplorchis* spp., *Centrocestus formosanus* and *Procerovum varium* which are endemic and infect people in many countries in Southeast Asia (Chai and Jung 2017). Moreover, the other genera/families, such as *Plagiorchis muris*, *Phaneropsolus bonnei*, *Fasciolopsis buski*, and *Gastrodiscoides hominis* are found to be widespread throughout many countries in Southeast Asia (Table 4.1). The remaining intestinal trematodes infecting people are rare and either have restricted endemic areas or their infection levels still remain underestimated.

### *4.3.2 Liver Flukes*

The liver flukes found in Southeast Asia consist of at least six species belonging to five genera and three families (Table 4.1). These liver flukes are *Opisthorchis viverrini* and *Clonorchis sinensis* which belong to the family Opisthorchiidae, two species of the genus *Fasciola*, i.e. *F. gigantica* and *F. hepatica* in the family Fasciolidae, and *Dicrocoelium dentriticum* and *Eurytrema pancreaticum* in the family Dicrocoeliidae (Table 4.1). While *E. pancreaticum* occurs mostly in pancreatic ducts, it is rarely found in liver bile ducts (Ishii et al. 1983).

*Opisthorchis viverrini* is the most medically important liver fluke causing significant public health problems in Southeast Asia. It is classified as group 1 carcinogen as it is the causative agent of liver and bile duct cancer (cholangiocarcinoma; CCA) in people. It is widespread throughout the lower Mekong Basin covering Thailand, Lao PDR, Cambodia and southern Vietnam, as well as Myanmar (Petney et al. 2018). At least 10 million people are currently infected in Southeast Asia, with the highest incidence of CCA in the world occurring in northeast Thailand (Petney et al. 2018). *Clonorchis sinensis* is also classified as a group 1 carcinogen, but in Southeast Asia it is restricted to the central-northern region of Vietnam (Doanh and Nawa 2016). These liver flukes are fish-borne and are transmitted by eating raw or partially cooked freshwater cyprinid fish (Grundy-Warr et al. 2012). However, there is a report of *O. viverrini*-like flukes infecting ducks in Vietnam, but there is no information of infection in people.

The genus *Fasciola* is originally a liver fluke of ruminants and consists of two principal species, *F. gigantica* and *F. hepatica* that also cause fascioliasis in people (Mas-Coma et al. 2019). The adult stage of these two species differs in size, which can be used to differentiate the species. Hybrid strains between these two species have been recorded as showing an intermediate morphological form and hence have been classified as *Fasciola* sp. (Mas-Coma et al. 2019). Fascioliasis is one of the most important zoonoses distributed worldwide, including Southeast Asia. Infection in people is caused by eating aquatic plants contaminated with infective metacercariae. Human fascioliasis has been reported at a quite low prevalence in Southeast Asia but there are no comprehensive reports on the current situation relating to infection in people in Myanmar, Thailand, or Vietnam (Tran et al. 2001).

*Dicrocoelium dentriticum* and *Eurytrema pancreaticum* mainly infect animals, thus infection of people, when it occurs, is accidental, rare, and restricted to some areas in the Philippines and Malaysia, respectively (Kumar 1999). People have also been found to be accidental hosts when they have eaten raw ants and grasshoppers, the second intermediate hosts of *D. dentriticum* and *E. pancreaticum*, respectively (Kumar 1999).

### *4.3.3 Lung Flukes*

There are currently at least 14 species of the lung flukes in genus *Paragonimus* that have been identified in Southeast Asia (Yoshida et al. 2019). Of these, four species *Paragonimus westermani*, *P. heterotremus*, *P. pseudoheterotremus*, and *P. philippinensis* have been reported to infect people causing paragonimiasis (Blair 2019). As the eggs of these lung flukes have morphological characteristics that are very similar, egg morphology alone cannot be used for species differentiation. Thus, species identification of lung flukes infecting people requires several methods, such as adult worm expulsion and specific serological or molecular diagnosis (Blair 2019).

The most common causative agent of paragonimiasis in people in Southeast Asia is *P. heterotremus* (Yoshida et al. 2019). Human paragonimiasis caused by *P. heterotremus* infection has been reported in Cambodia, Lao PDR, Myanmar, Thailand, and Vietnam (Yoshida et al. 2019). *P. heterotremus* infects people in at least 10 Provinces in Thailand (Kusolsuk et al. 2020). The most common paragonimiasis in Vietnam is also caused by *P. heterotremus* (Doanh et al. 2013). *Paragonimus westermani* causing human paragonimiasis in Southeast Asia has been diagnosed in the Philippines and Thailand (Blair 2019), and crabs containing the metacercariae of *P. westermani* were found in Cambodia, Lao PDR, Malaysia, Philippines, Thailand, and Vietnam (Blair 2019). Recently reported as a valid species, *P. pseudoheterotremus* (Waikagul 2007) has been detected in a patient's sputum in Thailand by molecular genotyping using the CO1 sequence (Intapan et al. 2012). However, due to the variation found in the metacercariae it is still controversial whether *P. heterotremus* and *P. pseudoheterotremus* are valid species or just represent geographical genetic variation within a single species (Doanh et al. 2015). *Paragonimus philippinensis* has recently been recovered from people and is found specifically in the Philippines (Ito et al. 1978).

### 4.3.4 Blood Flukes

There are three principal species belonging to the genus *Schistosoma* in Southeast Asia, namely *S. japonicum*, *S. mekongi,* and *S. malayensis*, which are the main causative agents of human schistosomiasis in this region (Gordon et al. 2019). Symptoms in people caused by schistosomiasis vary from hardly detectable to very serious. The most important human blood fluke in Southeast Asia is *S. japonicum*, which is still of considerable economic and public health concern in several areas of the Philippines and Indonesia, with an estimated 0.17–1.7% and 3% prevalence of infection, respectively (Zhou et al. 2010). Approximately 6.7 million people live in endemic areas of *S. japonicum* in the Philippines with 200,000 people estimated to be infected (Coutinho et al. 2005). While *S. japonicum* is currently endemic in three very isolated areas, namely Lindu, Napu, and Bada Valleys in Central Sulawesi Province, Indonesia, there is mounting evidence that the prevalence of *S. japonicum* infection is increasing in these areas (Satrija et al. 2015).

The second species is *Schistosoma mekongi* which has foci of distribution and transmission in the Mekong Delta (Gordon et al. 2019). The potential human population at risk from *S. mekongi* infection is currently estimated at over 1.5 million, with around 800 people infected in Lao PDR and around 2000 in Cambodia (Muth et al. 2010). The Khong and Mounlapamok districts in Lao PDR, and Ban Hat-Xai-Khong Island and San Dan, Sambour District in Kratie Province in Cambodia have recently been reported as remote areas of *S. mekongi* infection (Gordon et al. 2019). Transmission of *S. mekongi* has been discovered not only in tributaries of the Mekong river, but also within the Mekong Basin, namely at Sa Dao in the Xe Kong river of Cambodia (Attwood et al. 2004; Sinuon et al. 2007). Recent studies indicate that schistosomiasis has been emerging/re-emerging around Lake Inlay in central Myanmar. For instance, a schistosomiasis outbreak occurred in Rakhine State, Myanmar, with more than 400 confirmed cases and more than 800 suspected cases in 2018 (Gordon et al. 2019). Currently, the status of schistosomiasis in Myanmar is being assessed by molecular methods which have found that its causative agent is indeed *S. mekongi* (Wai et al. 2017).

*Schistosoma malayensis* was first described as *S. japonicum* from foreign nationals from China and Singapore living in Malaysia, with the later classification as *S. malayensis* (Chuah et al. 2019). To date, human infections in Malaysia appear to be restricted to West (peninsular) Malaysia and more specifically to the Jelai and Tembeling river systems that drain into the Pahang River in Pahang State. In general, cases of *S. malayensis* have been found in aboriginal Malaysians (Orang Asli) living in rural areas (Chuah et al. 2019). However, a low prevalence in people was observed, suggesting that people are not important hosts for this parasite, while *Rattus muelleri* and *Rattus tiomanicus* have been recorded as the main definitive hosts (Greer et al. 1988).

## 4.4 Diversity of Snail Intermediate Hosts

Snails are ubiquitous in freshwater bodies throughout tropical countries, especially in Southeast Asia, and a number of species are considered to be of medical importance since they serve as intermediate hosts of many trematode species (Fig. 4.3). Over 350 species of snails worldwide are estimated to be of possible medical or veterinary importance (Madsen and Hung 2014). Of these at least 45 species belonging to 26 genera of 7 families and are distributed within Southeast Asia where they act in the transmission of human trematodes (Table 4.2) (Madsen and Hung 2014; Lu et al. 2018).

At least nine and six species in two and four genera of snails occur in the families Ampullariidae and Viviparidae, respectively. These snails commonly act as intermediate hosts of intestinal flukes, such as the echinostomatids *A. malayanum*, *E. revolutum*, *E. ilocanum*, and *H. conoideum* (Madsen and Hung 2014; Lu et al. 2018). Moreover, the snails in these families are eaten as a common food in Southeast Asia, for example "goi hoi," which is prepared from raw apple snails or boiled snails (sometimes partially cooked) mixed in papaya salad called "tum hoi," or grilled snails known as "jee hoi" (Fig. 4.4). Even if these snails have been fried or boiled, the metacercariae of trematodes may not be completely destroyed. Studies have found that snails that have been boiled still contained infective metacercariae of *E. revolutum*, and that a high intensity of heat is needed with a cooking time of at least 19.2 min to completely destroy all metacercariae (Sulianti 2008). Thus, it is highly likely that inappropriate or undercooking of these snails in traditional dishes will contain infective metacercariae. Studies have found that the high rate of *E. revolutum* infection in children in Pursat Province, Cambodia, was due to these children eating partially cooked snails (see grilled snails in Fig. 4.4) after school from roadside stalls near their homes, and *E. ilocanum* was also recovered from the villagers in Oddar Meanchey Province, Cambodia (Sohn et al. 2011a, b).

There are three taxa of the genus *Bithynia*, family Bithyniidae, namely *Bithynia funiculata*, *B. siamensis goniomphalos*, and *B. s. siamensis* that act as the principal first intermediate hosts of the carcinogenic liver fluke, *O. viverrini*. In addition, cercariae and metacercariae of other trematodes, such as echinostomes, have been

**Table 4.2** The families and species of snails acting as intermediate hosts for trematode species

| Family | Genus and Species | Trematodes |
|---|---|---|
| Ampullariidae | *Pila ampullacea* | Echinostomes |
| | *Pila globosa* | Echinostomes |
| | *Pila gracilis* | Echinostomes |
| | *Pila luzonica* | Echinostomes |
| | *Pila pesmei* | Echinostomes |
| | *Pila polita* | Echinostomes |
| | *Pila scutata* | Echinostomes |
| | *Pomacea canaliculata* | Echinostomes |
| | *Pomacea maculata* | Echinostomes |
| Bithyniidae | *Bithynia funiculata* | *Opisthorchis viverrini*, echinostomes, heterophyids |
| | *Bithynia siamensis goniomphalos* | *O. viverrini*, echinostomes, heterophyid |
| | *Bithynia siamensis siamensis* | *O. viverrini*, *Clonorchis sinensis*, echinostomes, heterophyids |
| | *Parafossarulus manchouricus* | *C. sinensis* |
| | *Digoniostoma truncatum* | Heterophyids |
| | *Hydrobioides nassa* | Echinostomes |
| Lymnaeidae | *Austropeplea philippinensis* | *Fasciola* spp. |
| | *Austropeplea ollula* | *Fasciola* spp. |
| | *Bullastra cumingiana* | *Fasciola* spp. |
| | *Lymnaea (Radix) viridis* | *Fasciola* spp. |
| | *Lymnaea (Radix) auricularia* | *Fasciola* spp. |
| | *Lymnaea (Radix) swinhoei* | *Fasciola* spp. |
| | *Lymnaea (Radix) rubiginosa* | *Fasciola* spp., echinostomes |
| | *Radix quadrasi* | *Fasciola* spp. |
| Planorbidae | *Hippeutis (Helicorbis) umbilicalis* | *Fasciolopsis buski*, echinostomes |
| | *Segmentina (Polypylis) hemisphaerula* | *F. buski* |
| | *S. (Trochorbis) trochoideus* | *F. buski* |
| | *Indoplanorbis exustus* | Echinostomes |
| | *Gyraulus convexiusculus* | Echinostomes |
| | *Gyraulus sarasinorum* | Echinostomes |
| Pomatiopsidae | *Neotricula aperta* | *Schistosoma mekongi* |
| | *Oncomelania quadrasi* | *Schistosoma japonicum* |
| | *Oncomelania lindoensis* | *S. japonicum* |
| | *Robertsiella gismanni* | *S. japonicum* |
| Thiaridae | *Brotia asperata* | *Paragonimus* spp. |
| | *Brotia costula* | *Paragonimus* spp. |
| | *Melanoides tuberculata* | *Paragonimus* spp. |
| | *Sermyla riquetii* | *Paragonimus* spp. |

(continued)

**Table 4.2** (continued)

| Family | Genus and Species | Trematodes |
|---|---|---|
| | *Tarebia granifera* | *Paragonimus* spp. |
| | *Thiara scabra* | *Paragonimus* spp. |
| Viviparidae | *Bellamya javanica* | Echinostomes |
| | *Bellamya philippinensis* | Echinostomes |
| | *Filopaludina sumatrensis polygramma* | Echinostomes |
| | *Filopaludina martensi martensi* | Echinostomes |
| | *Taia polyzonata* | Echinostomes |
| | *Viviparus angularis* | Echinostomes |

**Fig. 4.4** Variety of dishes cooked from freshwater snails of the families Ampullariidae and Viviparidae, the risk trematode infection in Southeast Asia. (**a**) Papaya salad with snails or "tum hoi," (**b**) spicy snail salad or "goi hoi," (**c**) grilled snails or "jee hoi," (**d**) boiled snails or "tom hoi" (photos by Weerachai Saijuntha)

found in the same species of *Bithynia* snails (Bunchom et al. 2020). Other species of snails, for instance, *Digoniostoma truncatum* and *Hydrobioides nassa* are intermediate hosts of heterophyids and echinostomes, respectively (Madsen and Hung 2014; Bunchom et al. 2019). Furthermore, *B. s. siamensis* and *Parafossarulus manchouricus* in the family Bithyniidae have been reported as intermediate hosts of *C. sinensis* in Vietnam (Madsen and Hung 2014).

The family Lymnaeidae contains seven species in four genera that act as the principal intermediate hosts of fasciolid flukes, i.e. *Fasciola* spp. (Lu et al. 2018). Of these, the genus *Lymnaea* (*Radix*) is the most diverse and widespread in Southeast Asia (Madsen and Hung 2014), while the family Planorbidae comprises six species in four genera. Of these, three species, *Hippeutis* (*Helicorbis*) *umbilicalis*, *Segmentina* (*Polypylis*) *hemisphaerula*, and *Segmentina* (*Trochorbis*) *trochoideus* act as the principal intermediate hosts of *F. buski*, while the others *Indoplanorbis exustus*, *Gyraulus convexiusculus*, *G. sarasinorum* are intermediate hosts of echinostomatids (Table 4.2). The species of snails in these families are commonly found throughout Southeast Asia, especially in rice fields. After harvesting farmers release ducks to feed on the snails in their rice fields, which act to maintain and complete the life cycle of parasitic trematodes via a zoonotic cycle. A high prevalence of infection of echinostomes in free-grazing ducks has been reported in Thailand (Saijuntha et al. 2013).

The family Thiaridae consists of six species in five genera, which are the sole intermediate hosts of lung flukes in the genus *Paragonimus*. Of these, *Melanoides tuberculata* is the sole intermediate host for the most common causative agent of human paragonimiasis, *P. heterotremus*, which is widespread throughout Southeast Asia (Madsen and Hung 2014). The snails in the family Pomatiopsidae serve as the intermediate hosts of blood flukes, namely *Oncomelania quadrasi*, *O. lindoensis*, and *Robertsiella gismanni* acting as the intermediate hosts of *S. japonicum*, while *Neotricula aperta* is the sole intermediate host of *S. mekongi* (Madsen and Hung 2014; LoVerde 2019). A potential way to control and prevent schistosomiasis in remote areas is the eradication of snail intermediate hosts, which was successfully achieved in Japan, where schistosomiasis due to *S. japonicum* was eliminated by eradication of intermediate snail hosts, *Ocomelania* sp. (Takaka and Tsuji 1997).

## 4.5 Diversity of Second Intermediate Hosts and Sources of Transmission

Transmission of digenetic trematodes to people in Southeast Asia is almost entirely due to the regular consumption of food containing infective metacercariae. The variety of second intermediate hosts is a key to the successful maintenance, dispersal, and transmission of foodborne trematodes via foodborne transmission. There are aquatic animals and cyprinid species of fish acting as second intermediate hosts, as well as edible plants on which metacercariae are found, which are eaten as common, traditional daily meals in Southeast Asia. While waterborne transmission is known as the major transmission vehicle of schistosomiasis, some other trematodes can occasionally be transmitted via drinking untreated water from natural sources (Utzinger et al. 2010).

### 4.5.1 Foodborne Transmission

Freshwater plants harbor the metacercariae of the liver fluke genus *Fasciola* and the intestinal fluke genus *Fasciolopsis* which are the causative agents of fascioliasis and fasciolopsiasis, respectively, in people (Hillyer 1988). There are a range of edible freshwater plants that can be contaminated by the metacercariae of these flukes, such as morning glory, water mimosa, water fern, and gotu kola (Mas-Coma et al. 2018). In rural areas, these plants are usually harvested from natural puddles, then washed and eaten without heating. Thus, eating these plants without heating/boiling poses a high risk of *Fasciola* infection (Mas-Coma et al. 2018). In addition, an experimental study suggested that people consuming raw liver dishes from fresh livers infected with juvenile flukes could become infected resulting in subsequent fascioliasis (Taira et al. 1997; Mas-Coma et al. 2018). Moreover, the water chestnut, water caltrop, lotus, bamboo, and other edible fresh water plants have been shown to act as sources of infection for fasciolopsiasis (Grackyk et al. 2001).

A variety of aquatic invertebrates, such as snails, tadpoles, dragonfly naiads, crayfish, crabs, shrimps, and insect larvae act as the second intermediate hosts of foodborne trematodes in Southeast Asia. The lung fluke genus *Paragonimus* is transmitted by freshwater crabs or crayfish containing encysted metacercariae of *Paragonimus* spp. (Blair 2019). An alternative route of *Paragonimus* infection is the ingestion of raw meat from a mammalian paratenic host (Blair 2019). Intestinal flukes, however, are mainly transmitted via edible snails, dragonfly naiads, insect larvae, shrimps, and freshwater fish (Table 4.1). Local people usually eat these aquatic animals raw or partially cooked causing foodborne trematodiasis. In the case of the liver flukes *D. dentriticum* and *E. pancreaticum*, people have been found to be infected after accidentally eating raw ants and grasshoppers (Kumar 1999).

Cyprinid fish are the most diverse freshwater species acting as second intermediate hosts of trematodes. For instance, at least 60 cyprinid species (Table 4.3) have been found to be contaminated by the metacercariae of the liver flukes *O. viverrini* and *C. sinensis* in Southeast Asia (Saijuntha et al. 2019). Moreover, species of cyprinid fish are also the second intermediate hosts of several other trematodes, such as some species of intestinal flukes, heterophyids, and echinostomatids (Table 4.3). There are reports of a high prevalence of heterophyids, namely *Haplorchis* spp., *Centrocestus formosanus*, *Stellantchasmus falcatus* and echinostomatids, i.e. *Echinochasmus japonicus* infection in cyprinid fish from wild and fish-farming communities in Vietnam (Nguyen et al. 2007; Tran et al. 2008; Van et al. 2010), Thailand (Kumchoo et al. 2005), and Lao PDR (Eom et al. 2015). Therefore, consumption of raw or partially cooked cyprinid fish, including other

**Table 4.3** Cyprinid fish acting as intermediate hosts for liver and intestinal flukes with known distribution

| Cyprinid fish | Liver flukes infection | | Report on infection | Intestinal flukes infection | | Report on infection |
|---|---|---|---|---|---|---|
| | *O. viverrini* | *C. sinensis* | | Echinostomatids | Heterophyids | |
| *Amblyrhynchichthys truncatus* | ✔ | | Cambodia | | ✔ | Thailand |
| *Anabas testudineus* | | ✔ | Vietnam | | | |
| *Barbonymus altus* | ✔ | | Cambodia | | ✔ | Thailand |
| *Barbonymus gonionotus* | ✔ | | Thailand, Cambodia, Lao PDR | | ✔ | Thailand, Lao PDR |
| *Barbonymus schwanenfeldii* | ✔ | | Cambodia | | ✔ | Thailand |
| *Carassius auratus* | ✔ | | Vietnam | | ✔ | Vietnam |
| *Carassius cantonensis* | | ✔ | Vietnam | | | |
| *Carassius carassius* | | ✔ | Vietnam | | | |
| *Cirrhinus molitorella* | | ✔ | Vietnam | | ✔ | Vietnam |
| *Crossocheilus reticulatus* | ✔ | | Cambodia | | | |
| *Ctenopharyngodon idella* | | ✔ | Vietnam | | ✔ | Vietnam |
| *Culter recurvirostris* | | ✔ | Vietnam | | ✔ | Vietnam |
| *Cultrichthys erythropterus* | | ✔ | Vietnam | | | |
| *Cyclocheilichthys apogon* | ✔ | | Thailand, Cambodia, Lao PDR | | | |
| *Cyclocheilichthys armatus* | ✔ | | Thailand, Cambodia, Lao PDR | | | |
| *Cyclocheilichthys enoplos* | ✔ | | Cambodia, Lao PDR | | ✔ | Thailand |
| *Cyclocheilichthys furcatus* | ✔ | | Cambodia, Lao PDR | | | |
| *Cyclocheilichthys lagleri* | ✔ | | Cambodia | | | |
| *Cyclocheilichthys repasson* | ✔ | | Thailand, Cambodia, Lao PDR, Myanmar | | ✔ | Lao PDR |
| *Cyprinus carpio* | | ✔ | Vietnam | | ✔ | Vietnam |
| *Esomus metallicus* | ✔ | | Thailand, Lao PDR | | | |
| *Hampala dispar* | ✔ | | Thailand, Cambodia, Lao PDR | | ✔ | Thailand, Lao PDR |

| | | | | | | |
|---|---|---|---|---|---|---|
| *Hampala macrolepidota* | ✔ | | Thailand, Cambodia, Lao PDR | | ✔ | Thailand, Lao PDR |
| *Hemiculter leucisculus* | | ✔ | Vietnam | | ✔ | Vietnam |
| *Henicorhynchus lineatus* | ✔ | | Lao PDR | | | |
| *Henicorhynchus lobatus* | ✔ | | Cambodia | | | |
| *Henicorhynchus siamensis* | ✔ | | Thailand, Cambodia | | ✔ | Thailand |
| *Hypophthalmichthys molitrix* | | ✔ | Vietnam | | ✔ | Vietnam |
| *Hypsibarbus lagleri* | ✔ | | Lao PDR | | | |
| *Hypsibarbus pierrei* | ✔ | | Lao PDR | | | |
| *Hypsibarbus wetmorei* | ✔ | | Lao PDR | | ✔ | Thailand |
| *Labeo chrysophekadion* | ✔ | | Cambodia | | ✔ | Thailand |
| *Labeo rohita* | | ✔ | Vietnam | ✔ | ✔ | Vietnam |
| *Labiobarbus lineatus* | ✔ | | Thailand, Lao PDR | | | |
| *Labiobarbus siamensis* | ✔ | | Thailand, Cambodia | | ✔ | Thailand |
| *Lobocheilus melanotaenia* | ✔ | | Lao PDR | | | |
| *Mystacoleucus marginatus* | ✔ | | Lao PDR | | ✔ | Thailand |
| *Mylopharyngodon piceus* | | ✔ | Vietnam | | | |
| *Neolissochilus stracheyi* | ✔ | | Lao PDR | | | |
| *Onychostoma elongatum* | ✔ | | Lao PDR | | | |
| *Onychostoma fusiforme* | ✔ | | Lao PDR | | | |
| *Opsariichthys bidens* | | ✔ | Vietnam | | | |
| *Oreichthys parvus* | ✔ | | Lao PDR | | | |
| *Osteochilus hasseltii* | ✔ | | Thailand, Cambodia, Lao PDR | | ✔ | Thailand |
| *Osteochilus* sp. | ✔ | | Thailand | | ✔ | Thailand |
| *Osteochilus waandersii* | ✔ | | Lao PDR | | | |
| *Paralaubuca barroni* | ✔ | | Lao PDR | | | |
| *Poropuntius deauratus* | ✔ | | Lao PDR | | | |
| *Poropuntius laoensis* | ✔ | | Lao PDR | | | |

(continued)

**Table 4.3** (continued)

| Cyprinid fish | Liver flukes infection | | Report on infection | Intestinal flukes infection | | Report on infection |
|---|---|---|---|---|---|---|
| | *O. viverrini* | *C. sinensis* | | Echinostomatids | Heterophyids | |
| *Puntioplites falcifer* | ✔ | | Lao PDR | | | |
| *Puntioplites proctozystron* | ✔ | | Thailand, Cambodia, Lao PDR | | ✔ | Thailand |
| *Puntius brevis (Puntius leiacanthus)* | ✔ | | Thailand, Cambodia, Lao PDR, Myanmar, Vietnam | | | |
| *Puntius orphoides* | ✔ | | Thailand, Cambodia, Lao PDR | | | |
| *Puntius partipentazona* | ✔ | | Thailand | | | |
| *Puntius stoliczkanus* | ✔ | | Thailand | | | |
| *Puntius viehoever* | ✔ | | Thailand | | | |
| *Rasbora ourotacniatiran* | ✔ | | Lao PDR | | | |
| *Rasbora* sp. | ✔ | | Vietnam | | ✔ | Thailand, Vietnam |
| *Rhinogobius* sp. | | ✔ | Vietnam | | | |
| *Tilapia mossambica* | | ✔ | Vietnam | | | |
| *Thynnichthys thynnoides* | ✔ | | Thailand, Cambodia, | | | |
| *Toxabramis houdemeri* | | ✔ | Vietnam | | | |

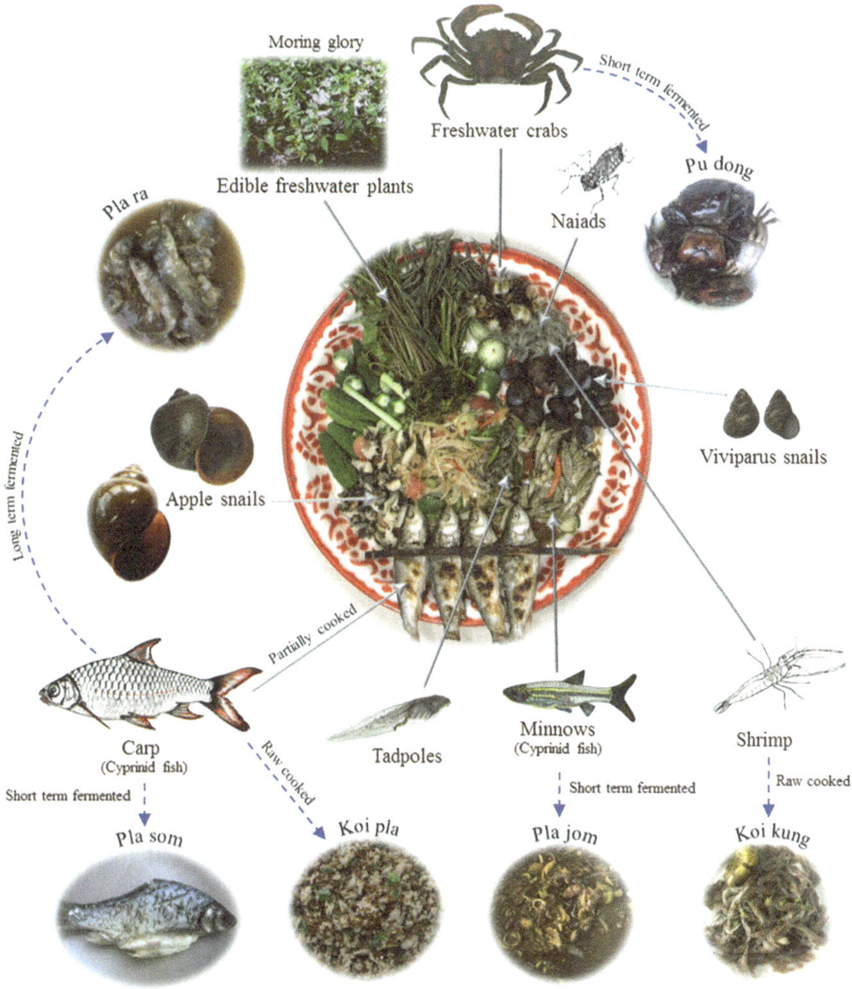


**Fig. 4.5** Sources of foodborne trematode infection via eating raw, partially cooked, or fermented edible aquatic plants and animals. This figure shows a favorite dish "som tum tad" or "papaya salad" in the northeast of Thailand style of cooking which frequently combines a variety of aquatic plants and animals (photo courtesy by Weerachai Saijuntha)

aquatic animals and edible aquatic plants are the major risk factors for trematode infection and subsequent disease in Southeast Asia (Fig. 4.5). Zoonotic trematodes have also been detected in raw fish dishes served in restaurants in Vietnam (Tran et al. 2009).

### 4.5.2 Waterborne Transmission

Schistosomiasis is predominantly transmitted via water contact. Local people who live in remote areas are at high risk of infection by drinking contaminated water. Larval forms of the blood fluke *Schistosoma* can penetrate into the skin and people subsequently develop schistosomiasis (LoVerde 2019). Not only the blood flukes, but some other trematodes are occasionally transmitted via water. For instance, metacercariae of *Fasciola*, *Fasciolopsis*, and *Paragonimus* can also encyst on the surface of water, and hence, people can be accidentally infected by drinking fresh untreated water contaminated with the metacercariae of these trematodes (Weng et al. 1980; Mas-Coma et al. 2018; Wang et al. 2008).

## 4.6 Genetic Diversity

A variety of molecular methods/markers have been used to examine the genetic diversity of human trematodes in Southeast Asia. Since the molecular era, multilocus enzyme electrophoresis (MEE) has been applied to investigate the systematics, genetic diversity, and population structure of several important human trematodes in Southeast Asia. In more recent times, DNA markers/techniques have been introduced to explore the phylogenetic relationships, genetic diversity and population genetics of these trematodes. Most information on the genetic diversity of human trematodes has been from comprehensive molecular/genetics studies, predominantly on the intestinal echinostomatids and heterophyids, the liver flukes *O. viverrini* and *C. sinensis*, lung flukes in the genus *Paragonimus*, and blood flukes in the genus *Schistosoma*. The following sections, therefore, will focus on the genetic diversity of these flukes.

### 4.6.1 Echinostomatids and Heterophyids

The taxonomy of these intestinal flukes, especially the echinostomatids in the family Echinostomatidae is still controversial as there are many synonyms represented in several species. Morphological characterization is often not sufficiently robust for accurate species identification; therefore, molecular genotyping has been used to provide independent molecular/genetic data for morphospecies differentiation. Genetic variation investigations of human echinostomes in the genera *Echinostoma*, *Artyfechinostomum* and *Hypoderaeum*, and heterophyids in the genus *Haplorchis* have been undertaken using MEE and several independent DNA markers.

Comprehensive studies on the systematics and genetic variation of echinostomatids in the family Echinostomatidae in Thailand and Lao PDR have been extensively conducted during the past 10 years. Genetic differentiation investigations between *Echinostoma* (*Artyfechinostomum*) *malayanum* and *Echinostoma revolutum* (Saijuntha et al. 2010a), and *E. revolutum* and *Hypoderaeum conoideum* (Saijuntha et al. 2010b) have been conducted by MEE, specifically allozyme electrophoresis. Subsequently, a series of studies were consecutively initiated, for instance, on investigations of genetic variation examining spatial and temporal populations of *E. revolutum* (Saijuntha et al. 2011a), and genetic variation and phylogenetic relationships of *E. revolutum*, *A. malayanum*, *H. conoideum*, and *Echinoparyphium recurvatum* (Saijuntha et al. 2011b). Later, the genetic differentiation between *A. malayanum* and *A. sulfratyfex* based on internal transcribed spacer sequences was examined (Tantrawatpan et al. 2013). Following these studies, mitochondrial DNA sequence analyses of 37 collar-spined echinostomes collected from domestic ducks in Thailand and Lao PDR revealed the presence of two species, *E. revolutum* and *E. miyagawai*, which were classified into genetic groups corresponding to the different continents, namely Eurasian (Europe and Asia), American, and Australian genetic groups (Nagataki et al. 2015).

In addition, the genetic variation and phylogenetic relationships of *H. conoideum* were examined by nuclear and mitochondrial sequence analyses (Tantrawatpan and Saijuntha 2020a). Most recently, nuclear intron sequence analyses revealed a degree of heterozygosity, which has provided potential genetic markers for population genetic studies of *E. revolutum* and *E. miyagawai* (Saijuntha et al. 2020). Based on ND1 sequence variation, multiplex PCR was successfully developed to differentiate *A. malayanum*, *E. ilocanum*, *E. revolutum*, and *H. conoideum* (Tantrawatpan and Saijuntha 2020b). The eggs of echinostomes are often morphologically similar to the eggs of *Fasciola*, *Fasciolopsis*, and *Paragonimus* flukes, which can lead to difficulties in distinguishing this diagnostic life cycle stage. Based on the magnitude of genetic differences between these flukes, several molecular techniques, such as real-time PCR, have been developed to accurately differentiate the eggs of these species (Tantrawatpan et al. 2016).

Genetic diversity investigations of heterophyids, *Haplorchis taichui* collected from people from Ha Giang, Thanh Hoa, and Quang Tri Provinces in Vietnam, have been reported. Based on mitochondrial CO1 sequence analysis, three genetic groups related to localities of collection have been confirmed by Dung et al. (2013). More recently, ribosomal transcription units have been used to demonstrate the phylogenetic identification of common heterophyids in Vietnam. A phylogenetic tree inferred from 28S rDNA sequences of trematodes clearly confirmed the status of each of the common species of Vietnamese heterophyids, namely *Centrocestus formosanus*, *Haplorchis pumilio*, *H. taichui*, *H. yokogawai*, *Procerovum varium*, and *Stellantchasmus falcatus* (Le et al. 2017). However, the eggs of heterophyid flukes are very similar morphologically to the liver fluke *O. viverrini,* thus several molecular techniques have been developed based on the genetic differences between these flukes, such as pyrosequencing to accurately differentiate eggs of *O. viverrini*, *C. sinensis*, *H. pumilio*, *H. taichui*, and *S. falcatus* (Tantrawatpan et al. 2014a).

### 4.6.2 *Genera* Opisthorchis *and* Clonorchis

Genetic diversity investigations of *O. viverrini* were initiated by MEE techniques using three enzyme loci to investigate the genetic variation of *O. viverrini* recovered from people (Sueblingvong et al. 1993). Subsequently, 33 enzyme loci (genetic markers) were established for genetic variation studies of *O. viverrini* by the MEE (allozyme) technique (Saijuntha et al. 2006). Of these, 32 enzyme loci were confidently used to study the systematics, genetic variation, and population genetics of *O. viverrini*. This demonstrated that *O. viverrini* has a population genetic sub-structure which is related to defined wetland (catchment) systems in Thailand and Lao PDR (Saijuntha et al. 2007). DNA markers/techniques, such as random amplified polymorphic DNA (RAPD), nuclear and mitochondrial DNA sequencing, microsatellite DNA including MEE have also been used for comprehensive genetic investigations of *O. viverrini* examining a variety of factors, such as spatial, temporal, and different host species (Sithithaworn et al. 2012; Petney et al. 2018). Based on such comprehensive studies, *O. viverrini* has now been defined as a "species complex" or "*O. viverrini* sensu lato" containing at least six genetically distinct groups correlated with five different wetland (catchment) systems in Thailand and Lao PDR (Saijuntha et al. 2007; Kiatsopit et al. 2014). Mitochondrial and nuclear DNA sequence variation of *O. viverrini* was also analyzed (Saijuntha et al. 2008; Pitaksakulrat et al. 2018). Subsequently, microsatellite markers were initially characterized (Laoprom et al. 2010) and then used for micro-scale population genetic investigations of *O. viverrini* populations in Khon Kaen Province, Thailand (Laoprom et al. 2012). Recently an additional novel genetic group from Sakon Nakhon Province, Thailand has been discovered by nuclear and mitochondrial DNA sequences including microsatellite DNA analyses (Namsanor et al. 2020).

Genetic diversity data for *C. sinensis* in Southeast Asia, particularly in Vietnam is very limited, with only a few studies that have been conducted using mitochondrial DNA sequences (Chelomina et al. 2014) and microsatellite DNA analyses (Nguyen et al. 2015). Mitochondrial CO1 sequencing has been used to explore the genetic diversity of *C. sinensis* within and between two geographical populations from Vietnam and Russia (Chelomina et al. 2014). There is a report using molecular detection that *C. sinensis* infects people in eastern Thailand (Traub et al. 2009), but later studies could not detect *C. sinensis* in the same geographical area (Buathong et al. 2017). More comprehensive studies on the systematics and genetic variation of *C. sinensis* in Vietnam and adjacent areas are urgently required. In addition, as the two species of liver flukes, *C. sinensis* and *O. viverrini* are endemic in Vietnam, multiplex PCR has been developed to differentiate these liver flukes based on their genetic differences (Le et al. 2006).

### *4.6.3 Genus* Fasciola

The liver flukes *Fasciola gigantica* and *F. hepatica*, including the intermediate (hybrid) form, *Fasciola* sp., have been genetically characterized in most countries in Southeast Asia. As morphological characters alone are often insufficient to accurately identify these fasciolid flukes, especially the egg stage as well as the hybrid form, molecular genetic techniques/markers have been developed for accurate genetic differentiation. For example, the PCR-RFLP patterns of ITS sequences can be used to differentiate the three fasciolid flukes (Ichikawa and Itagaki 2010). Moreover, there are several reports of aspermic *Fasciola* occurring in Southeast Asian countries (Itagaki et al. 2009; Ichikawa et al. 2011; Chaichanasak et al. 2012). PCR-RFLP of ITS1 revealed that aspermic *Fasciola* flukes in Thailand are identical to the *F. gigantica* genetic type (Chaichanasak et al. 2012), and this genetic type has been also found in Vietnam (Itagaki et al. 2009) and Myanmar (Ichikawa et al. 2011). Based on this molecular method a current survey of fasciolid flukes in Malaysia has found only *F. gigantic* types (Diyana et al. 2020), while both *F. gigantica* and *F. hepatica* genetic types have been found in the Philippines (Valino et al. 2017).

Several mitochondrial DNA sequences, however, have been used as genetic markers for intra-specific genetic variation investigations of the fasciolid flukes in Southeast Asia. Mitochondrial ND1 genotyping demonstrated that the aspermic fasciolid haplogroup was clearly genetically separated from the *F. gigantica* and *F. hepatica* haplogroups (Ichikawa et al. 2011; Chaichanasak et al. 2012). Moreover, *F. gigantica* was separated into Asian and African clades, while aspermic *Fasciola* clustered closely and was related to *F. gigantica* (Ichikawa et al. 2011; Nguyen et al. 2012). Based on genetic characterization studies it has been suggested that aspermic *Fasciola* may be expanding its distribution to non-endemic areas in Southeast Asia through the migration of domestic animals (Ichikawa et al. 2011; Chaichanasak et al. 2012). Most recently, novel DNA markers, such as the intron sequences of the taurocyamine kinase gene, have been characterized to explore genetic diversity and heterogeneity of the *Fasciola* flukes (Saijuntha et al. 2018). Intron sequence analyses revealed that DNA hybridization between *F. gigantica* and *F. hepatica* occurs and the hybrid form was identified. The study suggested that *Fasciola* sp. was closely related to both *F. gigantica* and *F. hepatica* based on the ratio of genetic materials of each species contained in the individual hybrid form (Saijuntha et al. 2018).

### *4.6.4 Genus* Paragonimus

Genetic variation of the lung flukes in the genus *Paragonimus*, which is endemic in Southeast Asia, has predominantly been undertaken by nuclear and mitochondrial DNA sequencing. These have been widely used for taxonomic and genetic diversity investigations (Blair et al. 2016). The most common species in the *P. heterotremus* complex have received the principal focus in studies examining genetic diversity in Southeast Asia. Early studies using a total of 18 enzymes (encoded by 20 loci) of *P. heterotremus* from Thailand were compared by MEE analyses with three Japanese species, *P. ohirai*, *P. miyazakii*, and *P. westermani*. This study found that *P. heterotremus* was more closely related to *P. miyazakii* than the other species (Agatsuma et al. 1992). The ITS2 and CO1 sequences have been used to investigate the genetic variation of *P. heterotremus* and *P. pseudoheterotremus* in Thailand and Myanmar. This study revealed that *P. heterotremus* from Thailand, Vietnam, and China formed a separate distinct phylogenetic clade. Metacercariae from Phitsanulok Province, Thailand, were found to be distinct from all others (Sanpool et al. 2013). However, the results indicated that *P. heterotremus* and *P. pseudoheterotremus* are not specifically genetically distinct from each other (Sanpool et al. 2013). This evidence was supported by a later study which showed that *P. heterotremus* and *P. pseudoheterotremus* isolates in Vietnam formed a *P. heterotremus* complex consisting of three genetic groups with strong geographical origins. According to these results, *P. heterotremus* and *P. pseudoheterotremus* should be considered geographically isolated populations of the *P. heterotremus* complex (Doanh et al. 2015). The results from these studies show that there is considerable genetic variation in the *P. heterotremus* complex in Southeast Asia.

Due to the genetic differences found in the CO1 sequence between *P. heterotremus* and *P. pseudoheterotremus*, a subsequent study of *P. pseudoheterotremus* in people in Thailand based on the CO1 sequence of DNA extraction from sputum showed that the sample was more similar to *P. pseudoheterotremus* (98–100%) than *P. heterotremus* (90–95%) (Intapan et al. 2012). Based on the DNA variation of the ITS2 region, a DNA pyrosequencing technique was successfully developed to differentiate 6 species of Southeast Asian *Paragonimus*, i.e. *Paragonimus bangkokensis*, *P. harinasutai*, *P. heterotremus*, *P. macrorchis*, *P. siamensis*, and *P. westermani* (Tantrawatpan et al. 2014b). Most recently the intron sequence of taurocyamine kinase has successfully been characterized and used to differentiate these six species, including *P. pseudoheterotremus*, as well as providing evidence of DNA hybridization between *P. heterotremus* and *P. pseudoheterotremus* (Tantrawatpan et al. 2021).

### *4.6.5 Genus* Schistosoma

Genetic diversity of *S. japonicum* in the Philippines and Indonesia has been comprehensively investigated based on geographical and host factors using mitochondrial DNA and microsatellite markers. Microsatellite markers have been isolated and characterized for *S. japonicum* and have shown that there is significant polymorphism between the different geographical isolates, making them highly useful for studying genetic diversity and the population genetic structure of this parasite (Shrivastava et al. 2003). For example, a study of mitochondrial and microsatellite markers of *S. japonicum* lineages from Indonesia, the Philippines, and Chinese

Taiwan has shown that the lineages were clearly distinct from each other and from those in mainland China (Yin et al. 2015). A further study using three mitochondrial gene sequences revealed that the isolates in the Philippines clustered closely to the isolates from Yunnan and Zhejiang Provinces in China, but were clearly distinct from the Japanese isolate (Chen et al. 2015). Microsatellite DNA analyses revealed a high level of transmission across human and animal host species, particularly between dogs and humans in the Philippines, which provides evidence for the role of dogs in transmission. This should be considered when planning and introducing control programs (Rudge et al. 2008). More recently, microsatellite DNA has been used to explore geographical strain differentiation between *S. japonicum* in the Philippines, where results suggested that geographical separation should be considered as one of the factors accounting for the observed differences between *S. japonicum* populations endemic in the Philippines (Moendeg et al. 2017).

Information on the genetic diversity of *S. mekongi* and *S. malayensis* is very limited, with only a few studies on DNA sequence information published. Phylogenetic analyses based on DNA sequences revealed that *S. malayensis* clustered as a sibling species of *S. mekongi* (Blair et al. 1997; Lockyer et al. 2003). Mitochondrial DNA sequence analyses of four populations of *S. mekongi* from Cambodia and Lao PDR showed that the three populations were distinguishable by mitochondrial loci (Attwood et al. 2008). Investigations on the genetic diversity within and among the *S. malayensis* and *S. mekongi* populations should be conducted and expanded to include more geographical isolates using additional polymorphic loci, such as microsatellite DNA, which has been successfully used for genetic diversity and population genetic analyses in the related species *S. japonicum*.

## 4.7 Conclusions

Human trematodes and their hosts in Southeast Asia are extremely diverse. Studies of their systematics, genetic diversity, and population genetics have commenced for a few species, but the majority of species infecting people in Southeast Asia still need more comprehensive, multidisciplinary studies. New species and new foci continue to be discovered and these need further research concerning their prevalence in the human population across their distributional ranges, as well as their life cycles, biology, and ecology. Species complexes of several trematodes have been detected, which now require comprehensive studies to define and understand their species level systematics, phylogenetic relationships, and population genetic structure. As there is a diverse range of edible plants contaminated with metacercariae and animals which act as intermediate hosts, studies should also examine the co-evolution or co-adaptation between trematodes and these plants and intermediate hosts and their definitive hosts. In conjunction with the above disciplines, biodiversity and genetic diversity investigations of host life cycles will provide essential information for the instigation and development of effective control and prevention programs against the trematodes that are endemic in Southeast Asia.

## References


Agatsuma T, Ketudat P, Thaithong S, Shibahara T, Sugiyama H, Habe S, Teraski K, Kawshima K (1992) Electrophoretic analysis of a natural population of the Thai *Paragonimus heterotremus* and its genetic relationship to the three Japanese species *P. miyazakii*, *P. ohirai* and *P. westermani*. Parasitol Res 78:463–468

Attwood SW, Campbell I, Upatham ES, Rollinson D (2004) Schistosomes in the Xe Kong River of Cambodia: the detection of *Schistosoma mekongi* in a natural population of snails and observations on the intermediate host's distribution. Ann Trop Med Parasitol 98:221–230

Attwood SW, Fatih FA, Upatham ES (2008) DNA-sequence variation among *Schistosoma mekongi* populations and related taxa; phylogeography and the current distribution of Asian schistosomiasis. PLoS Negl Trop Dis 2:e200

Blair D (2019) Paragonimiasis. Adv Exp Med Biol 1154:105–138

Blair D, van Herwerden L, Hirai H, Taguchi T, Habe S, Hirata M, Lai K, Upatham S, Agatsuma T (1997) Relationships between *Schistosoma malayensis* and other Asian schistosomes deduced from DNA sequences. Mol Biochem Parasitol 85:259–263

Blair D, Nawa Y, Mitreva M, Doanh PN (2016) Genetic diversity and genetic variation in lung flukes (genus *Paragonimus*). Trans R Soc Trop Med Hyg 110:6–12

Buathong S, Leelayoova S, Mungthin M, Ruang-areerate T, Naaglor T, Suwannahitatorn P, Piyaraj P, Taamasri P, Tan-ariya P (2017) Molecular discrimination of *Opisthorchis*-like eggs from residents in a rural community of central Thailand. PLoS Negl Trop Dis 11:e0006030

Bunchom N, Saijuntha W, Pilap W, Suksavate W, Vaisusuk K, Suganuma N, Agatsuma T, Petney TN, Tantrawatpan C (2019) Genetic variation of a freshwater snail *Hydrobioides nassa* (Gastropoda: Bithyniidae) in Thailand examined by mitochondrial DNA sequences. Hydrobiologia. https://doi.org/10.1007/s10750-019-04013-2

Bunchom N, Pilap W, Suksavate W, Vaisusuk K, Suganuma N, Agatsuma T, Saijuntha W (2020) Trematode infection in freshwater snails from Maha Sarakham Province, Thailand. Southeast Asian J Trop Med Public Health 51:518–527

Carbonell AC, Fried B (2019) Form and function in the digenea. Adv Exp Med Biol 1154:3–42

Chai JY, Jung BK (2017) Fishborne zoonotic heterophyid infection: an update. Food Waterborne Parasitol 8:33–63

Chai JY, Shin EH, Lee SH, Rim HJ (2009a) Foodborne intestinal flukes in Southeast Asia. Korean J Parasitol 47:S69–S101

Chai JY, Han ET, Shin EH, Sohn WM, Yong TS, Eom KS, Min DY, Um JY, Park MS, Hoang EH, Phommasack B, Insisiengmay B, Lee SH, Rim HJ (2009b) High prevalence of *Haplorchis taichui*, *Prosthodendrium molenkampi*, and other helminth infections among people in Khammouane Province, Lao PDR. Korean J Parasitol 47:243–247

Chai JY, Sohn WM, Yong TS, Eom KS, Min DY, Hoang EH, Phammasack B, Insisiengmay B, Rim HJ (2012) Echinostome flukes recovered from humans in Khammouane Province, Lao PDR. Korean J Parasitol 50:269–272

Chai JY, Sohn WM, Cho J, Eom KS, Yong TS, Min DY, Hoang EH, Phammasack B, Insisiengmay B, Rim HJ (2018) *Echinostoma ilocanum* infection in two residents of Savannakhet Province, Lao PDR. Korean J Parasitol 56:77–81

Chai JY, Cho J, Chang T, Jung BK, Sohn WM (2020) Taxonomy of *Echinostoma revolutum* and 37-collar-spined *Echinostoma* spp.: a historical review. Korean J Parasitol 58:343–371

Chaichanasak P, Ichikawa M, Sobhon P, Itagaki T (2012) Identification of *Fasciola* flukes in Thailand based on their spermatogenesis and nuclear ribosomal DNA, and their intraspecific relationship based on mitochondrial DNA. Parasitol Int 61:545–549

Chelomina GN, Tatonova YV, Hung NM, Ngo HD (2014) Genetic diversity of the Chinese liver fluke *Clonorchis sinensis* from Russia and Vietnam. Int J Parasitol 44:795–810

Chen F, Li J, Sugiyama H, Zhou DH, Song HQ, Zhao GH, Zhu XQ (2015) Genetic variability among *Schistosoma japonicum* isolates from the Philippines, Japan and China revealed by sequence analysis of three mitochondrial genes. Mitochondrial DNA 26:35–40

Cho J, Jung BK, Chang T, Sohn WM, Sinuon M, Chai JY (2020) Taxonomy of *Echinostoma revolutum* and 37-collar-spined *Echinostoma* spp.: a historical review. Korean J Parasitol 58:343–371
Chuah C, Gobert GN, Latif B, Heo CC, Leow CY (2019) Schistosomiasis in Malaysia: a review. Acta Trop 190:137–143
Coutinho HM, McGarvey ST, Acosta LP, Manalo DL, Langdon GC, Leenstra T, Kanzaria HK, Solomon J, Wu H, Olveda RM, Kurtis JD, Friedman JF (2005) Nutritional status and serum cytokine profiles in children, adolescents, and young adults with *Schistosoma japonicum*-associated hepatic fibrosis, in Leyte, Philippines. J Infect Dis 192:528–536
Diyana JNA, Mahiza MIN, Latiffah H, Fazila SH, Lokman IH, Hazfalinda HN, Chandrawathani P, Ibitoye EB, Juriah K (2020) Occurrence, morphometric, and molecular investigation of cattle and buffalo liver adult flukes in Peninsular Malaysia main abattoirs. J Parasitol Res 2020:5436846
Doanh PN, Nawa Y (2016) *Clonorchis sinensis* and *Opisthorchis* spp. in Vietnam: current status and prospects. Trans R Soc Trop Med Hyg 110:13–20
Doanh PN, Horii Y, Nawa Y (2013) *Paragonimus* and paragonimiasis in Vietnam: an update. Korean J Parasitol 51:621–627
Doanh PN, Thaenkham U, An PT, Hien HV, Horii Y, Nawa Y (2015) Metacercarial polymorphism and genetic variation of *Paragonimus heterotremus* (Digenea: Paragonimidae), and a re-appraisal of the taxonomic status of *Paragonimus pseudoheterotremus*. J Helminthol 89:182–188
Dung DT, Hop NT, Thaenkham U, Waikagul J (2013) Genetic differences among Vietnamese *Haplorchis taichui* populations using the COI genetic marker. J Helminthol 87:66–70
Eom KS, Park HS, Lee D, Sohn WM, Yong TS, Chai JY, Min DY, Rim HJ, Insisiengmay B, Phommasack B (2015) Infection status of zoonotic trematode metacercariae in fishes from Vientiane Municipality and Champasak Province in Lao PDR. Korean J Parasitol 53:447–453
Fried B, Graczyk TK, Tamang L (2004) Food-borne intestinal trematodiases in humans. Parasitol Res 93:159–170
Gordon CA, Kurscheid J, Williams GM, Clements ACA, Li Y, Zhou XN, Utzinger J, McManus DP, Gray DJ (2019) Asian schistosomiasis: current status and prospects for control leading to elimination. Trop Med Infect Dis 4:40
Grackyk TK, Gilman RH, Fried B (2001) Fasciolopsiasis: is it a controllable food-borne disease? Parasitol Res 87:80–83
Greer GJ, Ow-Yang CK, Yong HS (1988) *Schistosoma malayensis* n. sp.: a *Schistosoma japonicum*-complex schistosome from Peninsular Malaysia. J Parasitol 74:471–480
Grundy-Warr C, Andrews RH, Sithithaworn P, Petney TN, Sripa B, Laithavewat L, Ziegler AD (2012) Raw attitudes, wetland cultures, life-cycles: socio-cultural dynamics relating to *Opisthorchis viverrini* in the Mekong Basin. Parasitol Int 61:65–70
Hillyer GV (1988) Fascioliasis and fasciolopsiasis. In: Balows A, Hausler WJ, Ohashi M, Turano A, Lennete EH (eds) Laboratory diagnosis of infectious diseases. Springer, New York
Ichikawa M, Itagaki T (2010) Discrimination of the ITS1 types of *Fasciola* spp. based on a PCR-RFLP method. Parasitol Res 106:757–761
Ichikawa M, Bawn S, Maw NN, Htun LL, Thein M, Gyi A, Sunn K, Katakura K, Itagaki T (2011) Characterization of *Fasciola* spp. in Myanmar on the basis of spermatogenesis status and nuclear and mitochondrial DNA markers. Parasitol Int 60:474–479
Intapan PM, Sanpool O, Thanchomnang T, Imtawil K, Pongchaiyakul C, Nawa Y, Maleewong W (2012) Molecular identification of a case of *Paragonimus pseudoheterotremus* infection in Thailand. Am J Trop Med Hyg 87:706–709
Ishii Y, Koga M, Fujino T, Higo H, Ishibashi J, Oka K, Saito S (1983) Human infection with the pancreas fluke, *Eurytrema pancreaticum*. Am J Trop Med Hyg 32:1019–1022
Itagaki T, Sakaguchi K, Terasaki K, Sasaki O, Yoshihara S, Dung TV (2009) Occurrence of spermic diploid and aspermic triploid forms of *Fasciola* in Vietnam and their molecular characterization based on nuclear and mitochondrial DNA. Parasitol Int 58:81–85

Ito J, Yokogaw M, Araki K, Kobayashi M (1978) Studies on the morphology of larval and adult lung fluke in the Philippines, with a proposition of new name, *Paragonimus philippinensis*. Jpn J Parasitol 27:97–112

Johansen MV, Sithithaworn P, Bergquist R, Utzinger J (2010) Towards improved diagnosis of zoonotic trematode infection in Southeast Asia. Adv Parasitol 73:171–179

Kiatsopit N, Sithithaworn P, Saijuntha W, Petney TN, Andrews RH (2014) Analysis of population genetic of *Opisthorchis viverrini* sensu lato in Nam Ngum River wetland, Lao PDR by multilocus enzyme electrophoresis. Parasitol Res 11:2973–2981

Kostadinova A, Pérez-del-Olmo A (2019) The systematics of the trematode. Adv Exp Med Biol 1154:21–42

Kumar V (1999) Dicrocoeliasis and eurythremiasis. In: Trematode infections and diseases of man and animals. Springer, Dordrecht, pp 215–240

Kumchoo K, Wongsawad C, Chai JY, Vanittanakom P, Rojanapaibul A (2005) High prevalence of *Haplorchis taichui* metacercariae in cyprinoid fish from Chiang Mai Province, Thailand. Southeast Asian J Trop Med Public Health 36:451–455

Kusolsuk T, Phuphisut O, Maipanich W, Pubampen S, Sa-nguankiat S, Poodeepiyasawat A, Homsuwan N, Mongkolmoo S, Yoonuan T, Adisakwattana P, Silachamroon U, Nawa Y (2020) Asymtomatic human paragonimiasis among the Karen people in Tak Province, Thailand: a case report. Korean J Parasitol 58:57–60

Laoprom N, Sithithaworn P, Ando K, Sithithaworn J, Wongkham S, Laha T, Klinbugna S, Webster JP, Andrews RH (2010) Microsatellite loci in the carcinogenic liver fluke, *Opisthorchis viverrini* and their application as population genetic markers. Infect Genet Evol 10:146–153

Laoprom N, Sithithaworn P, Andrews RH, Ando K, Laha T, Klinbunga S, Webster JP, Petney TN (2012) Population genetic structure in *Opisthorchis viverrini* over various spatial scales in Thailand and Lao PDR. PLoS Negl Trop Dis 6:11

Le TH, De NV, Blair D, Sithithawom P, McManus DP (2006) *Clonorchis sinensis* and *Opisthorchis viverrini*: development of a mitochondrial-based multiplex PCR for their identification and discrimination. Exp Parasitol 112:109–114

Le TH, Nguyen KT, Nguyen NTB, Doan HTT, Dung DT, Blair D (2017) The ribosomal transcription units of *Haplorchis pumilio* and *H. taichui* and the use of 28S rDNA sequences for phylogenetic identification of common heterophyids in Vietnam. Parasite Vector 10:17

Lockyer AE, Olson PD, Østergaard P, Rollinson D, Johnston DA, Attwood SW, Southgate VR, Horak P, Snyder SD, Le TH, Agatsuma T, McManus DP, Carmichael AC, Naem S, Littlewood DTJ (2003) The phylogeny of the Schistosomatidae based on three genes with emphasis on the interrelationships of *Schistosoma* Weinland, 1858. Parasitology 126:203–224

LoVerde PT (2019) Schistostomiasis. Adv Exp Med Biol 1154:45–70

Lu XT, Gu QY, Limpanont Y, Song LG, Wu ZD, Okanurak K, Lv ZY (2018) Snail-borne parasitic diseases: an update on global epidemiological distribution, transmission interruption and control methods. Infect Dis Poverty 7:28

Madsen H, Hung NM (2014) An overview of freshwater snails in Asia with main focus on Vietnam. Acta Trop 140:105–117

Mas-Coma S, Bargues MD, Valero MA (2018) Human fascioliasis infection sources, their diversity, incidence factors, analytical methods and prevention measures. Parasitology 145:1665–1669

Mas-Coma S, Valero MA, Bargues MD (2019) Fascioliasis. Adv Exp Med Biol 1154:71–103

Moendeg KJ, Angeles JMM, Nakao R, Leonardo LR, Fontanilla KC, Goto Y, Kirinoki M, Villacorte EA, Rivera PT, Inoue N, Chigusa Y, Kawazu S (2017) Geographic strain differentiation of *Schistosoma japonicum* in the Philippines using microsatellite markers. PLoS Negl Trop Dis 11:e0005749

Muth S, Sayasone S, Odermatt-Biays S, Phompida S, Duong S, Odermatt P (2010) *Schistosoma mekogi* in Cambodia and Lao People's Democratic Republic. Adv Parasitol 72:179–203

Myers N, Mittermeier RA, Mittermeier CG, da Fonseca GAB, Kent J (2000) Biodiversity hostpots for conservation priorities. Nature 403:853–858

Nagataki M, Tantrawatpan C, Agatsuma T, Sugiura T, Duenngai K, Sithithaworn P, Andrews RH, Petney TN, Saijuntha W (2015) Mitochondrial DNA sequences of 37 collar-spined echinostomes (Digenea: Echinostomatidae) in Thailand and Lao PDR reveals presence of two species: *Echinostoma revolutum* and *E. miyagawai*. Infect Genet Evol 35:56–62

Namsanor J, Pitaksakulrat O, Kopolrat K, Kiatsopit N, Webster BL, Laha T, Saijuntha W, Laoprom N, Andrews RH, Petney TN, Sithithaworn P (2020) Two distinct genetic clusters of *Opisthorchis viverrini* are present in Sakon Nakhon Province, Thailand (Songkram River Basin): evidence from microsatellite and mitochondrial data. Int J Parasitol 6:11

Nguyen TD, Dalsgaard A, Thanh LLT, Murrell D (2007) Survey for zoonotic liver and intestinal trematode metacercariae in cluture and wild fish in an Giang Province, Vietnam. Korean J Parasitol 45:45–54

Nguyen S, Amer S, Ichikawa M, Itagaki T, Fukuda Y, Nakai Y (2012) Molecular identification of *Fasciola* spp. (Digenea: Platyhelmintes) in cattle from Vietnam. Parasite 19:85–89

Nguyen TTN, Arimatsu Y, Hong SJ, Brindley PJ, Blair D, Laha T, Sripa B (2015) Genome-wide characterization of microsatellites and marker development in the carcinogenic liver fluke *Clonorchis sinensis*. Parasitol Res 114:2263–2272

Petney TN, Andrews RH, Saijuntha W, Tesana S, Prasopdee S, Kiatsopit N, Sithithaworn P (2018) Taxonomy, ecology and population genetics of *Opisthorchis viverrini* and its intermediate hosts. Adv Parasitol 101:1–39

Pitaksakulrat O, Webster BL, Webster JP, Laha T, Saijuntha W, Lamberton P, Kiatsopit N, Andrews RH, Petney TN, Sithithaworn P (2018) Phylogenetic relationships within the *Opisthorchis viverrini* species complex with specific analysis of *O. viverrini* sensu lato from Sakon Nakhon, Thailand by mitochondrial and nuclear DNA sequencing. Infect Genet Evol 62:86–94

Rudge JW, Carabin H, Balolong E Jr, Tallo V, Shrivastava J, Lu DB, Basáñez MG, Olveda R, McGarvey ST, Webster JP (2008) Population genetics of *Schistosoma japonicum* within the Philippines suggest high levels of transmission between humans and dogs. PLoS Negl Trop Dis 2:e340

Saijuntha W, Sithithaworn P, Wongkham S, Laha T, Pipitgool V, Chilton NB, Petney TN, Andrews RH (2006) Genetic markers for the identification and characterization of *Opisthorchis viverrini*, a medically important food borne trematode in Southeast Asia. Acta Trop 100:246–251

Saijuntha W, Sithithaworn P, Wongkham S, Laha T, Pipitgool V, Tesana S, Chilton NB, Petney TN, Andrews RH (2007) Evidence of a species complex within the food-borne trematode *Opisthorchis viverrini* and possible co-evolution with their first intermediate hosts. Int J Parasitol 37:695–703

Saijuntha W, Sithithaworn P, Wongkham S, Laha T, Chilton NB, Petney TN, Andrews RH (2008) Mitochondrial DNA sequence variation among geographical isolates of *Opisthorchis viverrini* in Thailand and Lao PDR, and phylogenetic relationships with other trematodes. Parasitology 135:1479–1486

Saijuntha W, Tapdara S, Tantrawatpan C (2010a) Multilocus enzyme electrophoresis analysis of *Echinostoma revolutum* and *Echinostoma malayanum* (Trematoda: Echinostomatidae) isolated from Khon Kaen Province, Thailand. Asian Pac J Trop Med 3:633–636

Saijuntha W, Sithithaworn P, Andrews RH (2010b) Genetic differentiation of *Echinostoma revolutum* and *Hypodereaum conoideum* from domestic ducks in Thailand by multilocus enzyme electrophoresis. J Helminthol 84:143–148

Saijuntha W, Tantrawatpan C, Sithithaworn P, Andrews RH, Petney TN (2011a) Spatial and temporal genetic variation of *Echinostoma revolutum* (Trematoda: Echinostomatidae) from Thailand and the Lao PDR. Acta Trop 108:105–109

Saijuntha W, Sithithaworn P, Duenngai K, Kiatsopit N, Andrews RH, Petney TN (2011b) Genetic variation and relationships of four species of medically important echinostomes (Trematoda: Echinostomatidae) in South-East Asia. Infect Genet Evol 11:375–381

Saijuntha W, Deunngai K, Tantrawatpan C (2013) Zoonotic echinostome infection in free-grazing ducks in Thailand. Korean J Parasitol 51:663–667

Saijuntha W, Tantrawatpan C, Wang C, Agatsuma T, Intapan PM, Maleewong W, Petney TN (2018) Revealing genetic hybridization and DNA recombination of *Fasciola hepatica* and *Fasciola gigantica* in nuclear introns of the hybrid *Fasciola* flukes. Mol Biochem Parasitol 223:31–36

Saijuntha W, Sithithaworn P, Kiatsopit N, Andrews RH, Petney TN (2019) Liver flukes: *Clonorchis* and *Opisthorchis*. Adv Exp Med Biol 1154:139–180

Saijuntha W, Tantrawatpan C, Agatsuma T, Duenngai K, Sithithaworn P, Andrews RH, Petney TN (2020) Intron sequence variation of the echinostomes (Trematoda; Echinostomatidae): implications for genetic investigations of the 37 collar-spined, *Echinostoma miyagawai* Ischii, 1932 and *E. revolutum* (Fröelich, 1802). Parasitol Res 199:2485–2494

Sanpool O, Intapan PM, Thanchomnang T, Janwan P, Nawa Y, Maleewong W (2013) Molecular variation in the *Paragonimus heterotremus* complex in Thailand and Myanmar. Korean J Parasitol 6:677–681

Satrija F, Ridwan Y, Jsatal S, Rauf A (2015) Current status of schistosomiasis in Indonesia. Acta Trop 141:349–353

Shrivastava J, Barker G, Johansen MV, Xiaonong Z, Webster JP (2003) Isolation and characterisation of polymorphic DNA microsatellite markers from *Schistosoma japonicum*. Mol Ecol Notes 3:406–408

Sinuon M, Tsuyuoka R, Socheat D, Odermatt P, Ohmae H, Matsuda H, Montresor A, Palmer K (2007) Control of *Schistosoma mekongi* in Cambodia: results of eight years of control activities in the two endemic provinces. Trans R Soc Trop Med Hyg 101:34–39

Sithithaworn P, Andrews RH, Petney TN, Saijuntha W, Laoprom N (2012) The systematics and population genetics of *Opisthorchis viverrini sensu lato*: implications in parasite epidemiology and bile duct cancer. Parasitol Int 61:32–37

Sohn WM, Chai JY, Yong TS, Eom KS, Yoon CH, Sinuon M, Socheat D, Lee SH (2011a) *Echinostoma revolutum* infection in children, Pursat Province, Cambodia. Emerg Infect Dis 17:117–119

Sohn WM, Kim HJ, Yong TS, Eom KS, Jeong HG, Kim JK, Kang AR, Kim MR, Park JM, Ji SH, Sinuon M, Socheat D, Chai JY (2011b) *Echionstoma ilocanum* infection in Oddar Meanchey Province, Cambodia. Korean J Parasitol 49:187–190

Sueblingvong T, Piyamputra A, Priyanont S, Thaithong S (1993) Isoenzyme characterization of *Opisthorchis viverrini* form man. Chula Med J 37:375–385

Sulianti A (2008) Cooking *Echinostoma*-infected *Bellamya javanica* for safe consumption. Proc ASEAN Congr Trop Med Parasitol 3:28–32

Taira N, Yoshifuji H, Boray JC (1997) Zoonotic potential of infection with *Fasciola* spp. by consumption of freshly prepared raw liver containing immature flukes. Int J Parasitol 27:775–779

Takaka H, Tsuji M (1997) From discovery to eradication of schistosomiasis in Japan: 1874-1996. Int J Parasitol 27:1465–1480

Tantrawatpan C, Saijuntha W (2020a) Genetic variation and phylogenetic relationship of *Hypoderaeum conoideum* (Bloch, 1782) Dietz, 1909 (Trematoda: Echinostomatidae) inferred from nuclear and mitochondrial DNA sequences. Asian Pac J Trop Med 13:515–520

Tantrawatpan C, Saijuntha W (2020b) Multiplex PCR development for the differential detection of four medically important echinostomes (Trematoda: Echinostomatidae) in Thailand. Acta Trop 204:105304

Tantrawatpan C, Saijuntha W, Sithithaworn P, Andrews RH, Petney TN (2013) Genetic differentiation of *Artyfechinostomum malayanum* and *A. sufrartyfex* (Trematoda: Echinostomatidae) based on the internal transcribed spacer sequences. Parasitol Res 112:437–441

Tantrawatpan C, Intapan PM, Janwan P, Sanpool O, Lulitanond V, Srichantaratsamee C, Anamnart W, Maleewong W (2014a) Molecular identification of *Paragonimus* species by DNA pyrosequencing technology. Parasitol Int 62:341–345

Tantrawatpan C, Intapan PM, Thanchomnang T, Sanpool O, Janwan P, Lulitanond V, Sadaow L, Maleewong W (2014b) Development of a PCR assay and pyrosequencing for identification of important human fish-borne trematodes and its potential use for detection in fecal specimens. Parasit Vectors 7:88

Tantrawatpan C, Saijuntha W, Manochanrt S, Kheolamai P, Thanchomnang T, Sadaow L, Intapan PM, Maleewong W (2016) A singleplex real-time fluorescence resonance energy transfer PCR with melting curve analysis for the differential detection of *Paragonimus heterotremus*, *Echinostoma malayanum* and *Fasciola gigantica* eggs in faeces. Trans R Soc Trop Med Hyg 110:74–83

Tantrawatpan C, Tapdara S, Agatsuma T, Sanpool O, Intapan PM, Maleewong W, Saijuntha W (2021) Genetic differentiaion of Southeast Asian *Paragonimus* Braun, 1899 (Digenea: Paragonimidae) and genetic variation in the *Paragonimus heterotremus* complex examined by nuclear DNA sequences. Infect Genet Evol 90:104761

Toledo R, Esteban JG (2016) An update on human echinostomiasis. Trans R Soc Trop Med Hyg 110:37–45

Toledo R, Fried B (2019) Digenetic trematodes. Adv Exp Med Biol 1154:1–487

Toledo R, Alvárez-Izquierdo M, Muñoz-Antoli C, Esteban JG (2019) Intestinal trematode infection. Adv Exp Med Biol 1154:181–213

Tran VH, Tran TKD, Nguye HC, Pham HD, Pham TH (2001) Fascioliasis in Vietnam. Southeast Asian J Trop Med Public Health 32:48–50

Tran CTK, Dalsgaard A, Turnbull J, Tuan P, Murrell D (2008) Prevalence of zoonotic trematodes in fish from a Vietnamese fish farming community. J Parasitol 94:423–428

Tran CTK, Murrell D, Madsen H, Viet KN, Dalsgaard A (2009) Fishborne zoonotic trematodes in raw fish dishes served in restaurants in Nam Dinh Province and Hanoi, Vietnam. J Food Prot 72:2394–2399

Traub RJ, Macaranas J, Mungthin M, Leelayoova S, Cribb T, Murrell KD, Thompson RCA (2009) A new PCR-based approach indicates the range of *Clonorchis sinensis* now extends to central Thailand. PLoS Negl Trop Dis 3:e367

Utzinger J, Bergquist R, Olveda R, Zhou XN (2010) Important helminth infections in Southeast Asia: diversity, potential for control and prospects for elimination. Adv Parasitol 72:1–30

Valino LSI, Venturina VM, Mingala CN (2017) Gross and molecular comparison of *Fasciola hepatica* and *Fasciola gigantica* from the field in the Philippines. Int J Vet Sci 6:216–221

Van PT, Ersbøll AK, Nguyen TT, Viet KN, Nguyen HT, Murrell D, Dalsgaard A (2010) Freshwater aquaculture nurseries and infection of fish with zoonotic trematodes, Vietnam. Emerg Infect Dis 16:1905–1909

Wai KT, Han KT, Oo T (2017) Intensifying responsiveness towards neglected intestinal helminth infections in a resource-constrained setting. Trop Med Health 45:12

Waikagul J (2007) A new species of *Paragonimus* (Trematoda: Troglotrematidae) from a cat infected with metacercariae from mountain crabs *Larnaudia larnaudii*. J Parasitol 93:1496–1500

Wang CQ, Yang SJ, Pan HM, Wang DJ (2008) Study on *Paragonimus* infection by drinking water. Chin Prev Med 9:675–676

Weng YL, Zuang ZL, Lu CX, Yao TL (1980) Drinking water is a new transmission route of *Fasciolopsis buski* for human and animals. Chin J Zool 23:48–50

WHO (1995) Control of foodborne trematode infections. World Health Organ Tech Rep Ser 849:1–157

Yin M, Li H, McManus DP, Blair D, Su J, Yang Z, Xu B, Feng Z, Hu W (2015) Geographical genetic structure of *Schistosoma japonicum* revealed by analysis of mitochondrial DNA and microsatellite markers. Parasit Vectors 8:150

Yoshida A, Doanh PN, Maruyama H (2019) *Paragonimus* and paragonimiasis in Asia: an update. Acta Trop 199:105074

Zhou XN, Bergquist R, Leonardo L, Yang GJ, Yang K, Sudomo M, Olveda R (2010) Schistosomiasis japanica: control and research needs. Adv Parasitol 72:146–178

# Chapter 5
# The Community of Nematodes Inhabiting the Human Gut

**Oranuch Sanpool, Tongjit Thanchomnang, Hiroshi Yamasaki, Wanchai Maleewong, and Pewpan M. Intapan**

**Abstract** Southeast Asia is a tropical region in which a variety of parasitic diseases are endemic. Research into nematodes that inhabit the human gut is still in progress. However, there are sporadic gaps in the knowledge of species found in Southeast Asia, and there has been no attempt to extensively collate and integrate these data. This chapter reviews important highlights of the available information on nematodes that infect humans and that affect public health in Southeast Asia, with an emphasis on molecular identification and genetic diversity. This knowledge may be important for better understanding of nematodes that inhabit the human gut in these countries.

**Keywords** Human nematodes · *Strongyloides stercoralis* · Hookworms · *Capillaria philippinensis* · *Ascaris lumbricoides* · *Trichuris trichiura* · *Enterobius vermicularis*

## 5.1 Introduction

Gastrointestinal nematodes are the causative agents of parasitic diseases in the human body around the world and are mostly found in tropical countries where there is poor hygiene. Worldwide, more than 1.5 billion people are infected with soil-transmitted helminths (World Health Organization 2019). The effects of

O. Sanpool · W. Maleewong · P. M. Intapan (✉)
Department of Parasitology, Faculty of Medicine, Khon Kaen University, Khon Kaen, Thailand

Mekong Health Science Research Institute, Khon Kaen University, Khon Kaen, Thailand
e-mail: oransa@kku.ac.th; wanch_ma@kku.ac.th; pewpan@kku.ac.th

T. Thanchomnang
Faculty of Medicine, Mahasarakham University, Talat, Maha Sarakham, Thailand
e-mail: tongjit.t@msu.ac.th

H. Yamasaki
Department of Parasitology, National Institute of Infectious Diseases, Tokyo, Japan
e-mail: hyamasak@nih.go.jp



T. N. Petney et al. (eds.), *Biodiversity of Southeast Asian Parasites and Vectors causing Human Disease*, Parasitology Research Monographs 14,
https://doi.org/10.1007/978-3-030-71161-0_5

helminths on human health include abdominal disorder, anemia, malnutrition, and impairment of growth and development in young people. The gastrointestinal nematodes commonly found in humans living in Southeast Asian countries are soil-transmitted helminths such as *Ascaris lumbricoides*, *Trichuris trichiura*, *Trichostrongylus* species, *Strongyloides stercolaris*, and hookworms (*Necator americanus* and *Ancylostoma duodenale*). The routes via which these parasites infect the human body are generally divided into skin penetration by infective larvae and consumption of fresh food contaminated by nematode eggs. Moreover, nematode parasites that cause watery diarrhea, such as *Capillaria philippinensis,* are also found in the same places. Humans can become infected with *C. philippinensis* by eating raw cyprinid freshwater fish that contain infective larvae. People living in Southeast Asian countries prefer to eat traditionally fermented fish, which are undercooked; they do not like eating cooked fish.

The gold standards for diagnosis of nematode infection in the human gut are parasitological techniques for detection of parasite objects in fecal samples—i.e., eggs, larvae, or adults—by using a simple smear technique, the formalin–ether concentration technique, the Kato–Katz technique, etc. (World Health Organization 1991). However, these methods are time consuming and need experienced personnel. Molecular techniques are the most beneficial tools used for supportive diagnosis and for studying epidemiology and genetic diversity.

## 5.2 Nematodes That Infect the Human Small Intestine

### *5.2.1* Strongyloides *Species in Southeast Asian Countries (*Strongyloides stercoralis *and* Strongyloides fuelleborni*)*

#### 5.2.1.1 Parasite Distribution and Human Behavior Risk Factors for Infection

More than 50 species of *Strongyloides* are obligate gastrointestinal nematodes in vertebrates (mammals, birds, reptiles, and amphibians) (Speare 1989). Two main species—*S. stercoralis* and *S. fuelleborni*–like forms (including *S. fuelleborni kellyi*)—can infect humans. The parasites spread to humans by contact with soil. In endemic areas, infected persons defecate on the ground, where homogonic and heterogonic developments of the parasite take place and result in production of infective filariform larvae within a week. Rainfall helps spread infective larvae, but heavy rainfall resulting in flooding is detrimental to larval development because larvae are deprived of oxygen, which is essential for their growth and development (Anamnart et al. 2013). Human occupations involving soil contact are a predisposing factor for infection. Gardening in bare feet and farming activities, causing susceptibility to skin penetration by larvae, have been shown to be important risk factors (Senephansiri et al. 2017). Worldwide, strongyloidiasis is widely distributed, infecting 30–100 million people living in tropical and subtropical zones (Bethony et al. 2006). The overall estimated prevalence is between 10% and 40% among human populations in tropical and subtropical countries (Schär et al. 2013). Moreover, prevalence rates of up to 75% have also been found among refugees and immigrants in developing countries. Similar prevalence rates reported between 1992 and 2011 demonstrate that the global prevalence rates are as high as 50% in some areas of West Africa, the Caribbean, Southeast Asia, tropical regions of Brazil, Cambodia, and temperate regions of Spain, where there are humid soils and improper disposal of human feces. The prevalence increases between the age of <6 years and peaks in middle age, then it may decline (Forrer et al. 2018) or remain the same (Sithithaworn et al. 2003). Men who work in close contact with soil appear to have a higher prevalence than women (Wongsaroj et al. 2008; Forrer et al. 2018; Jongsuksuntigul et al. 2003).

*S. fuelleborni* is also prevalent in human populations in Africa and Southeast Asia. The first record of *S. fuelleborni* infection in humans was reported in Zimbabwe (Grove 1989), and infections were subsequently shown to be distributed widely in sub-Saharan countries (e.g., the Central African Republic, Cameroon, and Ethiopia) (Pampiglione and Ricciardi 1971; Hasegawa et al. 2010) and in Southeast Asia, including Thailand (Thanchomnang et al. 2017). This parasite is common among Old World primates, including Rhesus macaques (*Macaca* spp.) (Sandground 1925). In addition, *S. fuelleborni kellyi* has been reported in humans in Papua New Guinea (Ashford et al. 1992).

*S. fuelleborni*–like nematodes have been found in human feces in Papua New Guinea (Kelly et al. 1976). The worm was well described and was later found to cause swollen belly sickness in infants, which is occasionally fatal, among the Kamea people in Papua New Guinea (Vince et al. 2005). The worm was named *S. fuelleborni kellyi* after the first author of the 1976 publication (Ashford et al. 1992). A survey in children under 5 years of age revealed an infection rate of 27%, with varying intensity, as demonstrated by the fecal egg count (King and Mascie-Taylor 2004).

#### 5.2.1.2 Evidence and Identification of Strongyloidiasis in Humans

The prevalence of strongyloidiasis has been shown to be high in remote areas of Southeast Asia. With use of an agar plate culture (APC) method, rhabditiform and filariform larvae are detectable (Fig. 5.1). Surveys in northeastern Thai populations have shown prevalence rates of 23.5% (Jongsuksuntigul et al. 2003) and 28.9% (Sithithaworn et al. 2003), and a prevalence of 20.6% has been observed in southern Thailand (Wongsaroj et al. 2008). Other surveys have revealed prevalence rates of 48.6% among villagers in Cambodia (Forrer et al. 2018) and 41.0% in Laos (Laymanivong et al. 2016). In cases of autoinfection, rhabditiform larvae in the intestine molt rapidly in response to certain stimuli and develop into filariform larvae, which penetrate the intestinal wall. Autoinfection may explain the persistence of worms in chronic infections in humans (Gill et al. 2004). Definitive diagnosis

**Fig. 5.1** An agar plate stool culture survey for *Strongyloides* infection in a monkey. (**a**) A monkey living close to a human house in Laos. (**b**) The agar plate technique used for culturing *Strongyloides* at a survey site in Laos. (**c**) Various developmental stages of *S. fuelleborni* moving on an agar plate after 2 days of examination

relies on direct stool examination of eggs or larvae. Normally, the rhabditiform larvae of *S. stercoralis* are easy to identify under a microscope, but the small numbers of larvae in feces in asymptomatic and chronic cases can make them problematic to identify. Moreover, the appearance of larvae in stools fluctuates; thus, several samples are required to confirm the infection (Requena-Méndez et al. 2013). Conventional stool examination techniques used in health centers or hospitals include the direct simple smear, the cellophane thick smear, and the formalin–ether concentration technique (FECT), in which *Strongyloides* eggs (*S. fuelleborni*) or larvae (*S. stercoralis*) are detected and identified microscopically. Special stool techniques such as the Baermann method, Harada–Mori filter paper method, water emergence method, charcoal culture method, and agar plate culture method detect *Strongyloides* or hookworms in stools on the basis of development of filariform larvae in a soil environment and their crawling ability. Several methods include detection of antibodies or worm antigens in serum or other samples, but they await standardization and mass production to make them available to community hospitals and health centers worldwide. The sensitivities and specificities of both direct and indirect methods have been reviewed and summarized by Requena-Méndez et al. (2013). Molecular detection of larvae in stools has been developed to increase the sensitivity of detection. Indirect methods may be used in epidemiological surveys, as well as in diagnosis of strongyloidiasis.

Phylogenetic analysis of samples from community cross-sectional surveys in Laos has revealed parasite specimens belonging to *S. stercoralis*. Mitochondrial cytochrome *c* oxidase subunit I gene (*cox1*) sequencing has revealed that *S. stercoralis* has high diversity (24 haplotypes) (Laymanivong et al. 2016). Sequence analysis of 18S ribosomal RNA (18S rRNA) and *cox1* in *Strongyloides* samples from humans in Thailand has revealed that the species are *S. stercoralis* and *S. fuelleborni* (Thanchomnang et al. 2017). A median-joining network has shown

that the *S. stercoralis cox1* sequences fall into 43 known distinct haplotypes (Thanchomnang et al. 2017).

Polymerase chain reaction (PCR) is now widely used for detection of *Strongyloides* in biological samples. In an epidemiological survey of strongyloidiasis among school-aged children in northwestern Ethiopia, PCR revealed a higher prevalence (13.4%) than the Baermann and FECT techniques did (Amor et al. 2016). PCR can be used for follow-up analysis after ivermectin treatment (Repetto et al. 2018). Janwan et al. (2011) designed duplex real-time PCR to detect different intestinal parasites, which resulted in higher specificity and sensitivity than those of conventional parasitological methods. Nested PCR has also been developed, and its reported sensitivity is 100% (Sharifdini et al. 2015). In conclusion, molecular techniques not only enable detection of *Strongyloides* in specimens but also allow accurate identification of *Strongyloides* species, regardless of the development stage.

## 5.3 Genetic Diversity of Strongyloides and Related Species in Humans

*S. stercoralis* and *S. fuelleborni* were first reported in a human community that had contact with long-tailed macaques in Thailand (Thanchomnang et al. 2019). Moreover, dogs have been found to be reservoir hosts for possible transmission of human strongyloidiasis in Thailand (Sanpool et al. 2019). PCR-amplified 18S rRNA and *cox1* sequences have been used to identify the *Strongyloides* species present in human communities that have contact with long-tailed macaques in Thailand, and to assess the genetic diversity of the worms. The phylogenetic relationship of *S. stercoralis* and related species from Southeast Asia are shown in Fig. 5.2. *S. fuelleborni* has been genetically differentiated from *S. stercoralis*.

### *5.3.1 Hookworms and Related Roundworms (*Necator americanus, Ancylostoma duodenale, Ancylostoma ceylanicum, Trichostrongylus colubriformis, *and* Trichostrongylus axei*)*

#### 5.3.1.1 Parasite Distribution and Human Behavior Risk Factors for Infection

Hookworm infections in humans commonly cause socioeconomic and public health problems. Globally, the number of infected persons is approximately 1 billion (Schneider et al. 2011). Hookworm infections also causes iron deficiency anemia, resulting in mental retardation and growth insufficiency in children (Crompton 2000). *A. duodenale* and *N. americanus* are important species causing infection in

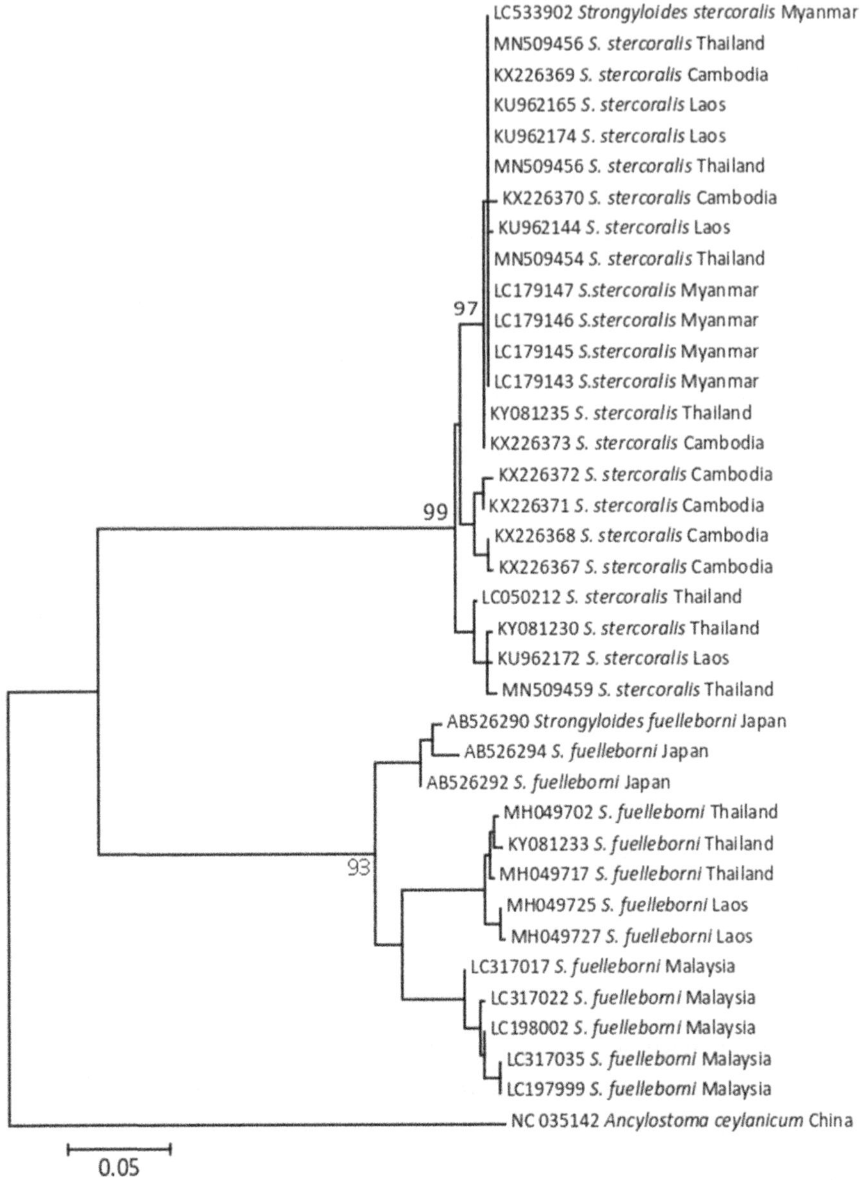


**Fig. 5.2** A maximum-likelihood phylogenetic tree based on partial mitochondrial cytochrome *c* oxidase subunit I gene (*cox1*) sequences (552 base pairs) of *Strongyloides stercoralis* and *S. fuelleborni.* Bootstrap scores (>90%, percentages of 1000 replications) are presented for each node. The sequences of *Strongyloides* species, obtained from the GenBank database, are indicated with their accession numbers and country name (Sanpool et al. 2019)

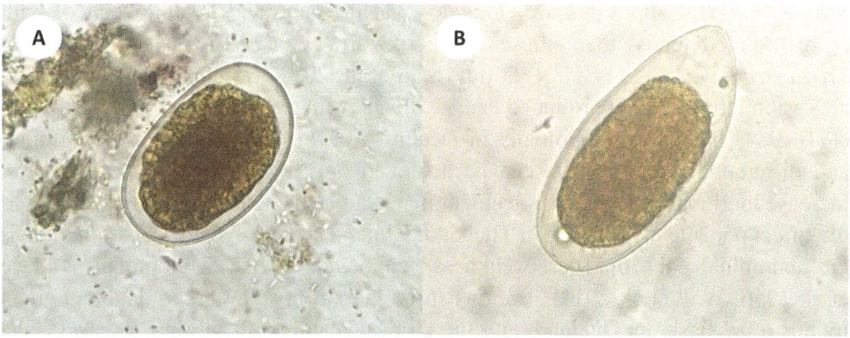


**Fig. 5.3** Eggs detected from stool examination. (**a**) A hookworm egg. (**b**) A *Trichostrongylus colubriformis* egg

humans (Chan et al. 1994). *A. duodenale* is commonly distributed in the Middle East, North Africa, India, Australia, and Europe, while *N. americanus* is widespread in Latin America and the Caribbean, sub-Saharan Africa, East Asia, and Southeast Asia (de Silva et al. 2003). In addition, zoonotic disease caused by *Ancylostoma* hookworms such as *A. ceylanicum, A. braziliense,* and *A. caninum* has been reported as a potentially significant public health problem in many areas (Mahdy et al. 2012).

*Trichostrongylus* species are common helminths in the digestive tract of herbivorous animals, especially livestock, and are widespread throughout the world (Yong et al. 2007). Most species occur in humans only by accidental infection, but a few pose considerable clinical and public health problems (Beaver et al. 1984). Human trichostrongyliasis cases have been reported sporadically in African countries and in many other countries, including Iran, Laos, Thailand, South Korea, China, the USA, and Australia (Ghadirian and Arfaa 1975; Beaver et al. 1984; Panasoponkul et al. 1985; Jariya et al. 1988; Boreham et al. 1995; Gutierrez et al. 2006; Sato et al. 2011), while *T. colubriformis* (Fig. 5.3) has been found in Thailand and Laos (Panasoponkul et al. 1985; Sato et al. 2011).

#### 5.3.1.2 Evidence and Identification of Hookworm and *Trichostrongylus* Infections in Humans

Although hookworm infections are still highly prevalent in Thailand (Jiraanankul et al. 2011; Jex et al. 2011), zoonotic hookworm disease caused by *A. ceylanicum* has been detected by copro-DNA methods in central parts of Thailand (Traub et al. 2008; Jiraanankul et al. 2011) and Laos (Sato et al. 2010).

*Trichostrongylus* adults are small nematodes (2–10 mm long by 50–80 μm wide), without distinct buccal capsules, which become embedded in the mucosa of the host's small intestine. Infection in humans is acquired through consumption of food or water contaminated with animal feces, which are used routinely as fertilizer. After reaching the small intestine, the larvae mature within 3–4 weeks and lay eggs.

*Trichostrongylus* spp. eggs are elongate–oval in shape and hyaline shelled, with more pointed ends than those of hookworm eggs. The average size of *Trichostrongylus* eggs is 85 × 45 μm (Sato et al. 2011), whereas hookworm eggs are oval but have broadly rounded ends, with an average size of 60 × 40 μm (Beaver et al. 1984). The egg morphologies of *Trichostrongylus* and hookworms are similar, and differential identification can be achieved only by parasitology experts. An increase in global population movement inside adjacent Asian countries may possibly affect parasite distribution in Thailand. Clinical symptoms of trichostrongyliasis are abdominal pain, diarrhea, weakness, leukocytosis, and eosinophilia (Wallace et al. 1956; Wall et al. 2011). There is still a lack of hospital-based data on human trichostrongyliasis in Thailand, and the clinical features of only a few cases have been presented in the literature (Panasoponkul et al. 1985; Jariya et al. 1988).

#### 5.3.1.3 Genetic Diversity of Hookworms and *Trichostrongylus* Species in Humans

Recently, molecular identification of hookworms and *Trichostrongylus* species found in infected humans in Southeast Asian countries such as Thailand, Laos, and Myanmar was reported (Phosuk et al. 2013a, 2013b; Aung et al. 2017). The species of hookworms recovered from humans were *A. duodenale*, *A. ceylanicum*, and *N. americanus*, and the *Trichostrongylus* species recovered were *T. colubriformis* and *T. axei*. The phylogenetic relationships between hookworms and *Trichostrongylus* species that infect populations in Southeast Asia has been evaluated (Fig. 5.4). These genetic data are important for epidemiological investigations and control (Phosuk et al. 2013a).

### *5.3.2* Capillaria philippinensis

#### 5.3.2.1 Parasite Distribution and Human Behavior Risk Factors for Infection

Intestinal capillariasis caused by *C. philippinensis*, a fish-borne nematode, is an important emerging zoonotic nematodiosis and has become more prevalent in recent years (McCarthy and Moore 2000). Fish encounter the infective eggs or larvae in water containing the feces of infected birds or humans. Humans are infected by consuming raw or undercooked small freshwater fish that the infective larvae are present. The parasite can reproduce within an individual host, resulting in a large number of worms in the small intestine. Consequently, if the parasite burden increases to massive levels, it can cause disease. Infections in humans were first recorded in the Philippines (Chitwood et al. 1968; Beaver et al. 1984). More cases have since been reported in the Philippines (Belizario et al. 2010), Indonesia (Chichino et al. 1992), Laos (Soukhathammavong et al. 2008), Japan (Cross

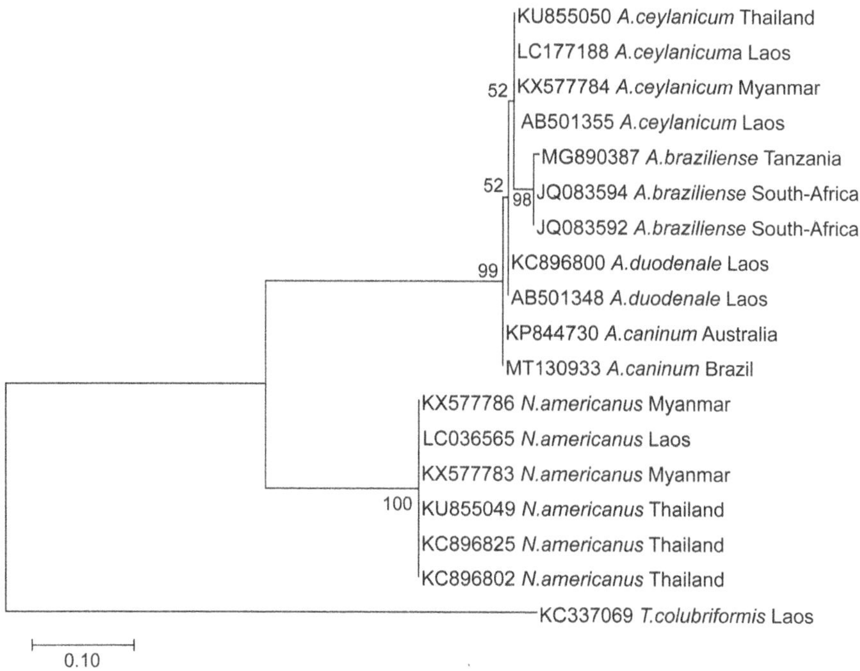


**Fig. 5.4** A maximum-likelihood phylogenetic tree based on the internal transcribed spacer 2 (ITS2) region of hookworm species, obtained from the GenBank database, with their accession numbers and country name

1992), Taiwan (Chen et al. 1989), India (Kang et al. 1994), Iran (Hoghooghi-Rad et al. 1987), and Egypt (Youssef et al. 1989; Mansour et al. 1990). There is also evidence that capillariasis has been transported to Europe from people's movement or migration an outside source (Chichino et al. 1992; Dronda et al. 1993; Austin et al. 1999). In Thailand, human cases have been reported in at least 25 provinces (Saichua et al. 2008). The first outbreak in Thailand was found in Sisaket Province in the northeast, where there were 20 cases with nine deaths (Kunaratanapruk et al. 1981).

#### 5.3.2.2 Evidence and Identification of Human Capillariasis

Human capillariasis is routinely diagnosed by examination of eggs, larvae, and/or adult *C. philippinensis* in the patient's feces. The egg is peanut shaped and light brown in color, with an average size of 36–45 × 21 μm (Fig. 5.5d). At the adult stage, both females and males are slender (capillary-like) in shape. The adult male is smaller than the female, and its posterior part has a long spicule covered with a sheath (Fig. 5.5a). The posterior half of the female's body contains a genital tract and vulva, which opens in the midbody, and a row of eggs can be seen in the uterus (Fig. 5.5c). Detection of antigens/antibodies by serological testing is used for

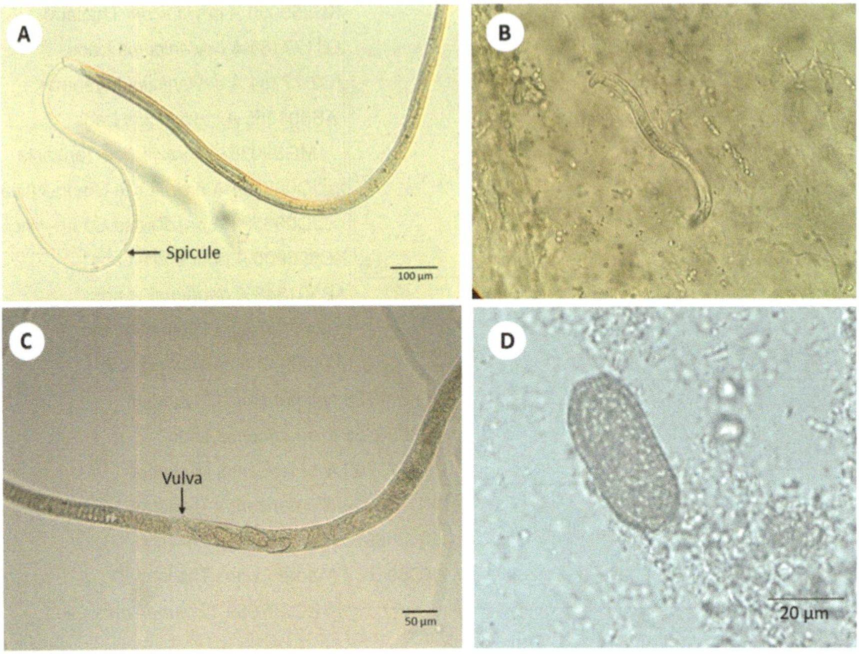


**Fig. 5.5** *Capillaria philippinensis* recovered from a human. (**a**) A male adult with a long spicule inside the posterior end; a spicular sheath is attached to the tail. (**b**) A larviparous larva found in a stool. (**c**) A female adult; the vulva opens in the midbody and a row of eggs is seen in the uterus. (**d**) The egg is peanut shaped, with a thick shell and flattened plugs at the bipolar ends

supportive diagnosis. An immunoblotting technique (Intapan et al. 2006) and a rapid diagnostic immunochromatographic device (Intapan et al. 2017) to detect antibodies in human sera are now available. Sadaow et al. (2018a) reported the clinical and epidemiological characteristics of 85 intestinal capillariasis patients in a hospital-based study in Thailand. The clinical manifestations—which included chronic diarrhea, borborygmus, abdominal pain, marked weight loss, muscle weakness, fatigue, dizziness, anorexia, and edema, as well as protein and electrolyte loss—are important for clinicians caring for patients in endemic areas to recognize. Infection could be avoided by expanded health communication from the responsible authorities. A public health control strategy is important to reduce the morbidity and mortality caused by this helminth disease in Thailand.

Molecular diagnosis using a specific nested PCR for detection of *C. philippinensis* in feces on the basis of amplification of small ribosomal subunits has been reported (El-Dib et al. 2015). This copro-DNA method can be useful for improved diagnosis and better understanding of transmission by detection of *C. philippinensis* DNA in intermediate hosts and reservoir hosts, and it can also help to improve strategies for surveillance and prevention of the disease.

#### 5.3.2.3 Prospective Genetic Diversity of *Capillaria philippinensis* and Related Species

There is still a lack of studies on the genetic diversity of *C. philippinensis* and related species in Southeast Asian countries. Molecular studies are need and would improve our understanding of the biology and transmission of *C. philippinensis.*

### *5.3.3* Ascaris lumbricoides

#### 5.3.3.1 Parasite Distribution and Human Behavior Risk Factors for Infection

*A. lumbricoides* is the largest nematode (roundworm) parasitizing the human intestine. It is one of the principal soil-transmitted helminths worldwide, infecting an estimated 819 million people (Holland 2013; Pullan et al. 2014). Ascariasis has been considered a neglected tropical disease by the World Health Organization and is very prevalent in rural areas and those with poor hygiene (Dold and Holland 2010).

#### 5.3.3.2 Evidence and Identification of Ascariasis in Humans

*Ascaris suum* is an intestinal roundworm found in pigs (Fig. 5.6a) but can cause human ascariasis (Arizono et al. 2010). *A. lumbricoides* and *A. suum* are

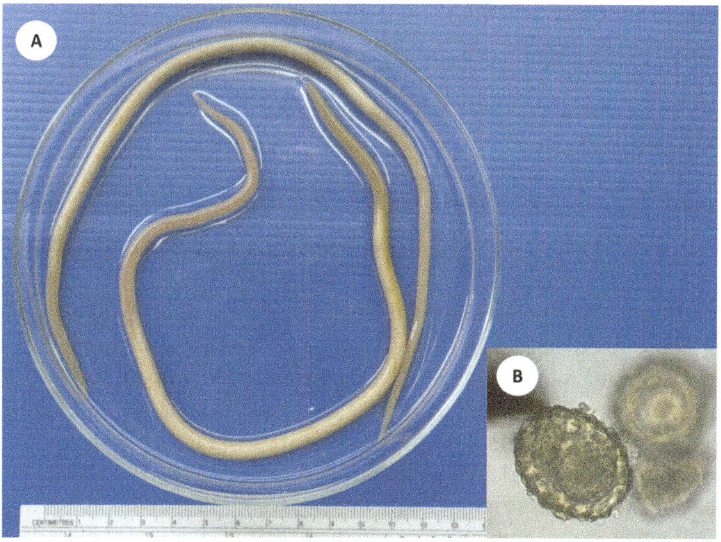


**Fig. 5.6** (**a**) An adult female *Ascaris suum* found in a pig. (**b**) Fertile eggs of an *Ascaris* species

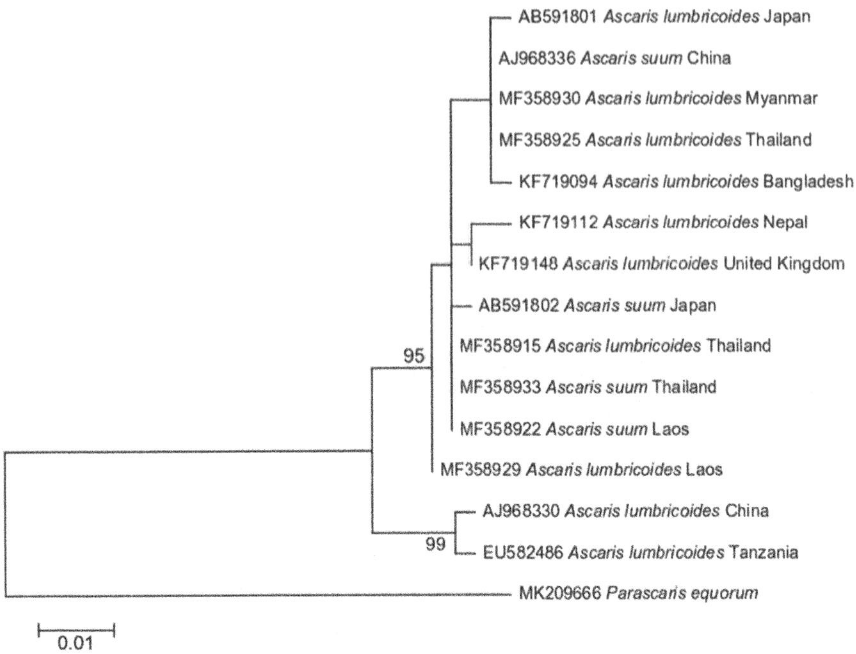


**Fig. 5.7** A maximum-likelihood phylogenetic tree based on partial mitochondrial cytochrome c oxidase subunit I gene (*cox1*) sequences of *Ascaris lumbricoides* and *A. suum*, obtained from the GenBank database, with their accession numbers and country name. Zoonotic cross-transmission occurs between pig and human ascariasis

morphologically similar and closely related (Fig. 5.7). There has been a debate as to whether they are distinct species; the final conclusion is that only a single species is involved (Leles et al. 2012; Shao et al. 2014). The reported prevalence rates of *Ascaris* infection in humans in Laos have ranged from 4.0% to 11.6% (Laymanivong et al. 2014, 2016).

#### 5.3.3.3 Genetic Diversity of *Ascaris lumbricoides* and Related Species

Several molecular–epidemiological investigations using polymorphic markers (internal transcribed spacer [ITS]-1, *cox1*, the NADH dehydrogenase subunit 1 gene (*nad1*), and microsatellite DNA markers) have explored the specificity of the two *Ascaris* species to their respective hosts and their taxonomic status (Leles et al. 2009; Iñiguez et al. 2012; Cavallero et al. 2013). Furthermore, knowledge on transmission dynamics and hybridization of *Ascaris* species has been gained by DNA sequencing, PCR restriction fragment length polymorphism (PCR-RFLP), and microsatellite marker techniques (Arizono et al. 2010; Cavallero et al. 2013; Betson et al. 2014; Jesudoss Chelladurai et al. 2017).

Recently, Sadaow et al. (2018b) reported the first molecular confirmation of *Ascaris* species in Thailand, Laos, and Myanmar. Sequence chromatograms of PCR-amplified ITS1 revealed a hybrid genotype from two human ascariasis cases in northern Thailand. All ITS2 sequences showed 100% identity with *A. lumbricoides* and *A. suum*. A phylogenetic tree of *cox1* sequences (Fig. 5.7) shows *A. lumbricoides* (which infects humans) and *A. suum* (which infects pigs) grouped in the same clade; thus, it is concluded that zoonotic cross-transmission between pigs and humans has occurred in these countries.

## 5.3.4 Trichuris trichiura *and Related Species*

### 5.3.4.1 Parasite Distribution and Human Behavior Risk Factors for Infection

Human trichuriasis, caused by *T. trichiura*, is a soil-transmitted helminth infection. It is grouped among the important neglected tropical diseases by the World Health Organization (Hotez et al. 2009; Dunn et al. 2016). An estimated 600–800 million people are infected worldwide (Bethony et al. 2006; Dunn et al. 2016). Various *Trichuris* species have been found in mammalian hosts: *T. suis* in swine, *T. vulpis* in canines, *T. ovis* in sheep, *T. skrjabini* in goats, and *T. muris* in mice (Liu et al. 2013). Among these, it is thought that *T. vulpis* (Márquez-Navarro et al. 2012) and *T. suis* (Kradin et al. 2006; Liu et al. 2014) can establish persistent active infections in humans. *T. trichiura* and *T. suis* adults can be discriminated on the basis of morphological parameters and biometrical determinations (Cutillas et al. 2009). However, differentiation of these species by egg morphology is difficult.

### 5.3.4.2 Evidence and Identification of Trichuriasis in Humans

Human trichuriasis is usually asymptomatic and can present with mild symptoms. Persons with heavy infections can experience diffuse colitis, chronic diarrhea, abdominal cramps, rectal tenesmus, and rectal prolapse (Dunn et al. 2016; Stephenson et al. 2000), some of which have important health consequences (Lenk et al. 2016). The prevalence of human trichuriasis is high in Central African countries, South India, and Southeast Asia (de Silva et al. 2003), and it can be high in children (Stephenson et al. 2000), particularly schoolchildren (aged 5–14 years) (Azira and Zeehaida 2012). In the lower Greater Mekong Subregion, the following prevalence rates of trichuriasis have been reported: 8.5% in Laos (Laymanivong et al. 2014), 57% in Myanmar (Montresor et al. 2004), 4.1% in Cambodia (Yong et al. 2014), and 6.3% in pregnant women in southern Thailand (Liabsuetrakul et al. 2009). Trichuriasis is diagnosed routinely by identification of *Trichuris* eggs in fecal specimens (Ok et al. 2009). The eggs of different species of *Trichuris* are very similar, making it difficult to differentiate between species on the basis of egg morphology (Cutillas et al. 2007) (Fig. 5.8).

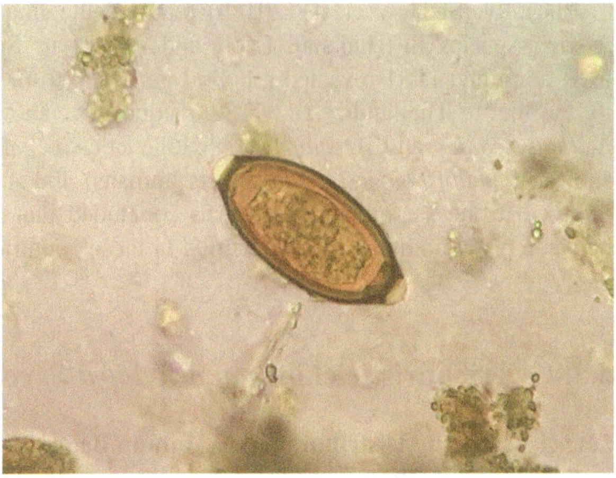

**Fig. 5.8** An unembryonated *Trichuris trichiura* egg in human feces. It is barrel shaped and thick shelled, with a mucoid plug at each end, and is 50–55 μm long by 20–25 μm wide

Molecular techniques are increasingly being used as supportive tools for identification at the species level (Liu et al. 2013, 2014). Several molecular markers are useful for identification of *Trichuris* spp., including nucleotide sequences of ITS1, ITS2, nuclear 18S rRNA, and *cox1* (Cutillas et al. 2007; Doležalová et al. 2015; Meekums et al. 2015).

#### 5.3.4.3 Genetic Diversity of *Trichuris trichiura* and Related Species

Discrimination of closely related *Trichuris* species has been reported by Callejón et al. (2013). Recently, molecular identification of the species of *Trichuris* eggs collected from human fecal samples in Laos, Myanmar, and Thailand has been reported (Phosuk et al. 2018). A phylogenetic tree based on the ITS2 sequences of *Trichuris* species and related species is shown in Fig. 5.9. The molecular evidence of *T. trichiura* and *T. suis* infection in humans is clear, and greater awareness of the zoonotic potential of *T. suis* infection is needed in Southeast Asian countries. Molecular systematic, taxonomic, and diagnostic studies in human populations associated with *T. trichiura* and *T. suis* are important for epidemiological investigations.

### *5.3.5* Enterobius vermicularis

#### 5.3.5.1 Parasite Distribution and Human Behavior Risk Factors for Infection

*E. vermicularis*, a pinworm, is a global helminth parasite causing enterobiasis in humans. It is especially common in temperate climates. Approximately 4–28% of

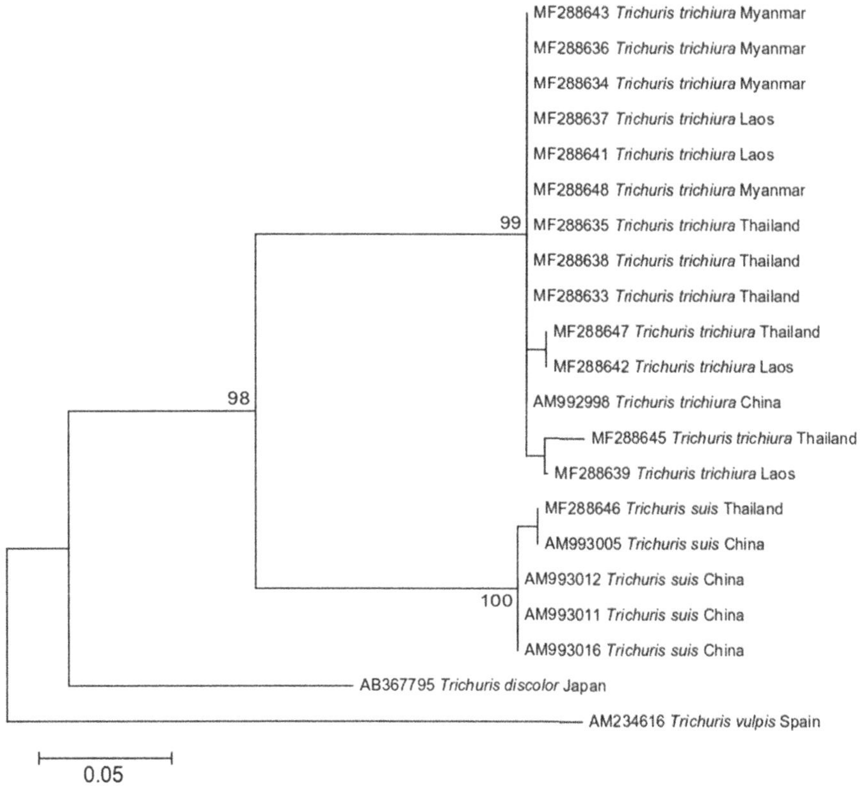


**Fig. 5.9** A maximum-likelihood phylogenetic tree based on partial internal transcribed spacer 2 (ITS2) sequences of *Trichuris trichiura* and *T. suis* found in humans, clearly showing their close relationship and the zoonotic potential of *T. trichiura* and *T. suis* infections in human hosts when placed in context with other *Trichuris* species

children are infected worldwide (Bethony et al. 2006). In Thailand, the prevalence of *E. vermicularis* infection among primary schoolchildren ranges from 0.19% to 38.82% (Jongsuksantigul et al. 1992; Kaewkes et al. 1983; Nithikathkul et al. 2001; Polseela et al. 2004; Saksirisampant et al. 2006). The prevalence is generally related to local levels of public health support, education, and personal hygiene (Li et al. 2015). Human enterobiasis is also closely associated with high population density, socioeconomic status, and the infant habit of thumb sucking (Kim et al. 2013). In addition, reinfection with enterobiasis occurs commonly by the fecal–oral route.

#### 5.3.5.2 Evidence and Identification of Enterobiasis in Humans

The parasitological method used for *E. vermicularis* diagnosis is microscopic identification of eggs collected in the perianal area. *E. vermicularis* eggs are transparent, elongate–oval in shape, and slightly flattened on one side (Fig. 5.10). They are usually embryonated when shed. The presence of adult worms is also diagnostic when they are found in the perianal area. Female worms emerge in the perianal area to lay eggs at night, resulting in perianal itching and irritation (Cerva et al. 1991). Eggs become infective within a few hours of being laid and can survive for several days on clothing or bed linen.

#### 5.3.5.3 Genetic Diversity of *Enterobius vermicularis* and Related Species

Primeval evidence of *E. vermicularis* has been found. Parasite DNA from pre-Columbian human coprolites, dating back 6110 years, has been sequenced and found to be very similar to that of extant *E. vermicularis* (Iñiguez et al. 2006). In Thailand, Tomanakan et al. (2018) reported the first molecular identification of *E. vermicularis* from schoolchildren and documented genetic variation among eggs by using sequence analyses of *cox1* and ITS2. All ITS2 sequences were identified as *E. vermicularis*. Phylogenetic analysis of *E. vermicularis cox1* sequences shows 66 haplotypes: six haplotypes from Thailand are in cluster A, with sequences from Japan and Korea, and five haplotypes are in cluster B, with sequences from Japan, Iran, Poland, Greece, and Denmark (Nakano et al. 2006). However, no sequence from a Thai haplotype is in cluster C of *E. vermicularis*, which has been found in captive chimpanzees in Japan. The phylogenetic tree shows high genetic diversity among *E. vermicularis* sequences, with closely related transmission in humans and primates (Fig. 5.11).

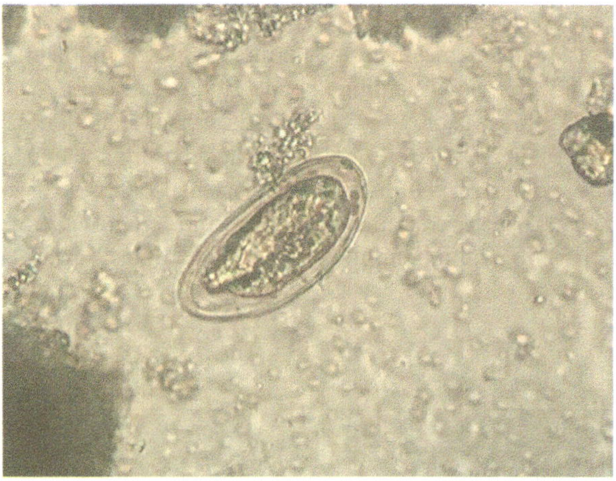

**Fig. 5.10** An *Enterobius vermicularis* egg. It is characteristically asymmetrical about the long axis, being distinctly flattened on one side and elongate–oval in shape. An infective larva is seen inside the egg

**Fig. 5.11** A maximum-likelihood phylogenetic tree based on partial mitochondrial cytochrome *c* oxidase subunit I gene (*cox1*) sequences of *Enterobius vermicularis*. This is a new construction from the GenBank database, modified on the basis of a report by Tomanakan et al. (2018). The

**Acknowledgements** The authors Oranuch Sanpool, Tongjit Thanchomnang, Wanchai Maleewong, and Pewpan Intapan were supported by a Distinguished Research Professor grant (number DPG6280002) from the Thailand Research Fund.

## References

Amor A, Rodriguez E, Saugar JM, Arroyo A, López-Quintana B, Abera B et al (2016) High prevalence of *Strongyloides stercoralis* in school-aged children in a rural highland of north-western Ethiopia: the role of intensive diagnostic work-up. Parasit Vectors 9(1):617

Anamnart W, Pattanawongsa A, Intapan PM, Morakote N, Janwan P, Maleewong W (2013) Detrimental effect of water submersion of stools on development of *Strongyloides stercoralis*. PLoS One 8(12):e82339

Arizono N, Yoshimura Y, Tohzaka N, Yamada M, Tegoshi T, Onishi K et al (2010) Ascariasis in Japan: is pig-derived *Ascaris* infecting humans? Jpn J Infect Dis 63(6):447–448

Ashford RW, Barnish G, Viney ME (1992) *Strongyloides fuelleborni kellyi*: infection and disease in Papua New Guinea. Parasitol Today 8(9):314–318

Aung WPP, Htoon TT, Tin HH, Sanpool O, Jongthawin J, Sadaow L et al (2017) First molecular identifications of *Necator americanus* and *Ancylostoma ceylanicum* infecting rural communities in lower Myanmar. Am J Trop Med Hyg 96(1):214–216

Austin DN, Mikhail MG, Chiodini PL, Murray-Lyon IM (1999) Intestinal capillariasis acquired in Egypt. Eur J Gastroenterol Hepatol 11(8):935–936

Azira NMS, Zeehaida M (2012) Severe chronic iron deficiency anaemia secondary to *Trichuris* dysentery syndrome—a case report. Trop Biomed 29(4):626–331

Beaver PC, Jung RC, Cupp EW (1984) Aphasmid nematodes. In: Beaver PC, Jung RC, Cupp EW (eds) Clinical parasitology, 9th edn. Lea & Febiger, Philadelphia, pp 231–252

Belizario VY Jr, Totañes FIG, de Leon WU, Migriño JR, Macasaet LY (2010) Intestinal capillariasis, western Mindanao, the Philippines. Emerg Infect Dis 16(4):736–738

Bethony J, Brooker S, Albonico M, Geiger SM, Loukas A, Diemert D et al (2006) Soil-transmitted helminth infections: ascariasis, trichuriasis, and hookworm. Lancet 367(9521):1521–1532

Betson M, Nejsum P, Bendall RP, Deb RM, Stothard JR (2014) Molecular epidemiology of ascariasis: a global perspective on the transmission dynamics of *Ascaris* in people and pigs. J Infect Dis 210(6):932–941

Boreham RE, McCowan MJ, Ryan AE, Allworth AM, Robson JM (1995) Human trichostrongyliasis in Queensland. Pathology 27(2):182–185

Callejón R, Nadler S, De Rojas M, Zurita A, Petrášová J, Cutillas C (2013) Molecular characterization and phylogeny of whipworm nematodes inferred from DNA sequences of *cox1* mtDNA and 18S rDNA. Parasitol Res 112(11):3933–3949

Cavallero S, Snabel V, Pacella F, Perrone V, D'Amelio S (2013) Phylogeographical studies of *Ascaris* spp. based on ribosomal and mitochondrial DNA sequences. PLoS Negl Trop Dis 7(4): e2170

Cerva L, Schrottenbaum M, Kliment V (1991) Intestinal parasites: a study of human appendices. Folia Parasitol 38(1):5–9

---

**Fig. 5.11** (continued) haplotypes from Thailand (labeled with *red stars*) are in cluster A (found in East Asia) and in cluster B (found in Denmark, Germany, Greece, and Poland [in Europe]; Iran [in the Middle East]; and Japan [in East Asia], where *E. vermicularis* is found in captive chimpanzees) (Nakano et al. 2006). Other haplotypes of *E. vermicularis* found in captive chimpanzees in Japan are in cluster C. *E. anthropopitheci* is an outer species

Chan MS, Medley GF, Jamison D, Bundy DA (1994) The evaluation of potential global morbidity attributable to intestinal nematode infections. Parasitology 109(Pt 3):373–387
Chen CY, Hsieh WC, Lin JT, Liu MC (1989) Intestinal capillariasis: report of a case. Taiwan Yi Xue Hui Za Zhi 88(6):617–620
Chichino G, Bernuzzi AM, Bruno A, Cevini C, Atzori C, Malfitano A et al (1992) Intestinal capillariasis (*Capillaria philippinensis*) acquired in Indonesia: a case report. Am J Trop Med Hyg 47(1):10–12
Chitwood MB, Valesquez C, Salazar NG (1968) *Capillaria philippinensis* sp. n. (Nematoda: Trichinellida), from the intestine of man in the Philippines. J Parasitol 54(2):368–371
Crompton DW (2000) The public health importance of hookworm disease. Parasitology 121 (Suppl):S39–S50
Cross JH (1992) Intestinal capillariasis. Clin Microbiol Rev 5(2):120–129
Cutillas C, de Rojas M, Ariza C, Ubeda JM, Guevara D (2007) Molecular identification of *Trichuris vulpis* and *Trichuris suis* isolated from different hosts. Parasitol Res 100(2):383–389
Cutillas C, Callejón R, de Rojas M, Tewes B, Ubeda JM, Ariza C et al (2009) *Trichuris suis* and *Trichuris trichiura* are different nematode species. Acta Trop 111(3):299–307
de Silva NR, Brooker S, Hotez PJ, Montresor A, Engels D, Savioli L (2003) Soil-transmitted helminth infections: updating the global picture. Trends Parasitol 19(12):547–551
Dold C, Holland CV (2010) *Ascaris* and ascariasis. Microbes Infect 13:632–637
Doležalová J, Oborník M, Hajdušková E, Jirků M, Petrželková KJ, Bolechová P et al (2015) How many species of whipworms do we share? Whipworms from man and other primates form two phylogenetic lineages. Folia Parasitol 62:63
Dronda F, Chaves F, Sanz A, Lopez-Velez R (1993) Human intestinal capillariasis in an area of nonendemicity: case report and review. Clin Infect Dis 17(5):909–912
Dunn JC, Turner HC, Tun A, Anderson RM (2016) Epidemiological surveys of, and research on, soil-transmitted helminths in Southeast Asia: a systematic review. Parasit Vectors 9:31
El-Dib NA, El-Badry AA, Ta-Tang TH, Rubio JM (2015) Molecular detection of *Capillaria philippinensis*: an emerging zoonosis in Egypt. Exp Parasitol 154:127–133
Forrer A, Khieu V, Schär F, Vounatsou P, Chammartin F, Marti H et al (2018) *Strongyloides stercoralis* and hookworm co-infection: spatial distribution and determinants in Preah Vihear Province, Cambodia. Parasit Vectors 11(1):33
Ghadirian E, Arfaa F (1975) Present status of trichostrongyliasis in Iran. Am J Trop Med Hyg 24(6):935–941
Gill GV, Welch E, Bailey JW, Bell DR, Beeching NJ (2004) Chronic *Strongyloides stercoralis* infection in former British Far East prisoners of war. QJM 97(12):789–795
Grove DI (1989) Clinical manifestations. In: Grove DI (ed) Strongyloidiasis—a major roundworm infection of man. Taylor & Francis, London, pp 155–173
Gutierrez Y, Guerrant RL, Walker DH, Weller PF (2006) Other tissue nematode infections. In: Guerrant RL, Walker DH, Weller PF (eds) Tropical infectious diseases, principles, pathogens and practice, 2nd edn. Churchill Livingstone, Oxford, pp 1231–1247
Hasegawa H, Sato H, Fujita S, Nguema PP, Nobusue K, Miyagi K et al (2010) Molecular identification of the causative agent of human strongyloidiasis acquired in Tanzania: dispersal and diversity of *Strongyloides* spp. and their hosts. Parasitol Int 59(3):407–413
Hoghooghi-Rad N, Maraghi S, Narenj-Zadeh A (1987) *Capillaria philippinensis* infection in Khoozestan Province, Iran: case report. Am J Trop Med Hyg 37(1):135–137
Holland C (2013) *Ascaris*: the neglected parasite. Elsevier, Amsterdam
Hotez PJ, Fenwick A, Savioli L, Molyneux DH (2009) Rescuing the bottom billion through control of neglected tropical diseases. Lancet 373(9674):1570–1575
Iñiguez AM, Reinhard K, Gonçalves MLC, Ferreira LF, Araújo A, Vicente ACP (2006) SL1 RNA gene recovery from *Enterobius vermicularis* ancient DNA in pre-Columbian human coprolites. Int J Parasitol 36(13):1419–1425
Iñiguez AM, Leles D, Jaeger LH, Carvalho-Costa FA, Araújo A, Amazonas Research Group (2012) Genetic characterisation and molecular epidemiology of *Ascaris* spp. from humans and pigs in Brazil. Trans R Soc Trop Med Hyg 106(10):604–612

Intapan PM, Maleewong W, Sukeepaisarnjaroen W, Morakote N (2006) Potential use of *Trichinella spiralis* antigen for serodiagnosis of human capillariasis philippinensis by immunoblot analysis. Parasitol Res 98(3):227–231

Intapan PM, Rodpai R, Sanpool O, Thanchomnang T, Sadaow L, Phosuk I et al (2017) Development and evaluation of a rapid diagnostic immunochromatographic device to detect antibodies in sera from intestinal capillariasis cases. Parasitol Res 116(9):2443–2447

Janwan P, Intapan PM, Thanchomnang T, Lulitanond V, Anamnart W, Maleewong W (2011) Rapid detection of *Opisthorchis viverrini* and *Strongyloides stercoralis* in human fecal samples using a duplex real-time PCR and melting curve analysis. Parasitol Res 109(6):1593–1601

Jariya P, Lertlaituan P, Junnoo V, Bedavanij A, Tesjaroen S (1988) One human case of trichostrongyliasis in Thailand. J Med Assoc Thai 71(8):469–470

Jesudoss Chelladurai J, Murphy K, Snobl T, Bader C, West C, Thompson K et al (2017) Molecular epidemiology of *Ascaris* infection among pigs in Iowa. J Infect Dis 215(1):131–138

Jex AR, Lim YA, Bethony JM, Hotez PJ, Young ND, Gasser RB (2011) Soil-transmitted helminths of humans in Southeast Asia—towards integrated control. Adv Parasitol 74:231–265

Jiraanankul V, Aphijirawat W, Mungthin M, Khositnithikul R, Rangsin R, Traub RJ et al (2011) Incidence and risk factors of hookworm infection in a rural community of central Thailand. Am J Trop Med Hyg 84(4):594–598

Jongsuksuntigul P, Chaeychomsri W, Techamontrikul P (1992) Study on prevalence and intensity of intestinal helminthiasis and opisthorchiasis in Thailand. J Trop Med Parasitol 15:80–95

Jongsuksuntigul P, Intapan PM, Wongsaroj T, Nilpan S, Singthong S, Veerakul S et al (2003) Prevalence of *Strongyloides stercoralis* infection in northeastern Thailand (agar plate culture detection). J Med Assoc Thai 86(9):737–741

Kaewkes S, Tesana S, Sithithaworn P, Srisawangwonk T, Raengsangounwong P (1983) Enterobiasis in young school children in Bangkok. J Trop Med Parasitol 6:19–24

Kang G, Mathan M, Ramakrishna BS, Mathai E, Sarada V (1994) Human intestinal capillariasis: first report from India. Trans R Soc Trop Med Hyg 88(2):204

Kelly A, Little MD, Voge M (1976) *Strongyloides fuelleborni*-like infections in man in Papua New Guinea. Am J Trop Med Hyg 25(5):694–699

Kim DH, Cho MK, Park MK, Kang SA, Kim BY, Park SK et al (2013) Environmental factors related to enterobiasis in a southeast region of Korea. Korean J Parasitol 51(1):139–142

King SE, Mascie-Taylor CGN (2004) *Strongyloides fuelleborni kellyi* and other intestinal helminths in children from Papua New Guinea: associations with nutritional status and socioeconomic factors. P N G Med J 47(3–4):181–191

Kradin RL, Badizadegan K, Auluck P, Korzenik J, Lauwers GY (2006) Iatrogenic *Trichuris suis* infection in a patient with Crohn disease. Arch Pathol Lab Med 130(5):718–720

Kunaratanapruk S, Iam-Ong S, Chatsirimongkol C (1981) Intestinal capillariasis: the first endemic in Thailand. Ramathibodi Med J 4:209–213

Laymanivong S, Hangvanthong B, Keokhamphavanh B, Phommasansak M, Phinmaland B, Sanpool O et al (2014) Current status of human hookworm infections, ascariasis, trichuriasis, schistosomiasis mekongi and other trematodiases in Lao People's Democratic Republic. Am J Trop Med Hyg 90(4):667–669

Laymanivong S, Hangvanthong B, Insisiengmay B, Vanisaveth V, Laxachack P, Jongthawin J et al (2016) First molecular identification and report of genetic diversity of *Strongyloides stercoralis*, a current major soil-transmitted helminth in humans from Lao People's Democratic Republic. Parasitol Res 115(8):2973–2980

Leles D, Araújo A, Vicente AC, Iñiguez AM (2009) Molecular diagnosis of ascariasis from human feces and description of a new *Ascaris* sp. genotype in Brazil. Vet Parasitol 163(1–2):167–170

Leles D, Gardner SL, Reinhard K, Iñiguez A, Araujo A (2012) Are *Ascaris lumbricoides* and *Ascaris suum* a single species? Parasit Vectors 5:42

Lenk EJ, Redekop WK, Luyendijk M, Rijnsburger AJ, Severens JL (2016) Productivity loss related to neglected tropical diseases eligible for preventive chemotherapy: a systematic literature review. PLoS Negl Trop Dis 10(2):e0004397

Li HM, Zhou CH, Li ZS, Deng ZH, Ruan CW, Zhang QM et al (2015) Risk factors for *Enterobius vermicularis* infection in children in Gaozhou, Guangdong, China. Infect Dis Poverty 4:28
Liabsuetrakul T, Chaikongkeit P, Korviwattanagarn S, Petrueng C, Chaiya S, Hanvattanakul C et al (2009) Epidemiology and the effect of treatment of soil-transmitted helminthiasis in pregnant women in southern Thailand. Southeast Asian J Trop Med Public Health 40(2):211–222
Liu GH, Gasser RB, Nejsum P, Wang Y, Chen Q, Song HQ et al (2013) Mitochondrial and nuclear ribosomal DNA evidence supports the existence of a new *Trichuris* species in the endangered françois' leaf-monkey. PLoS One 8(6):e66249
Liu GH, Zhou W, Nisbet AJ, Xu MJ, Zhou DH, Zhao GH et al (2014) Characterization of *Trichuris trichiura* from humans and *T. suis* from pigs in China using internal transcribed spacers of nuclear ribosomal DNA. J Helminthol 88(1):64–68
Mahdy MA, Lim YA, Ngui R, Siti Fatimah MR, Choy SH, Yap NJ et al (2012) Prevalence and zoonotic potential of canine hookworms in Malaysia. Parasit Vectors 5:88
Mansour NS, Anis MH, Mikhail EM (1990) Human intestinal capillariasis in Egypt. Trans R Soc Trop Med Hyg 84(1):114
Márquez-Navarro A, García-Bracamontes G, Alvarez-Fernández BE, Ávila-Caballero LP, Santos-Aranda I, Díaz-Chiguer DL et al (2012) *Trichuris vulpis* (Froelich, 1789) infection in a child: a case report. Korean J Parasitol 50(1):69–71
McCarthy J, Moore TA (2000) Emerging helminth zoonoses. Int J Parasitol 30(12–13):1351–1360
Meekums H, Hawash MBF, Sparks AM, Oviedo Y, Sandoval C, Chico ME et al (2015) A genetic analysis of *Trichuris trichiura* and *Trichuris suis* from Ecuador. Parasit Vectors 19(8):168
Montresor A, Zin TT, Padmasiri E, Allen H, Savioli L (2004) Soil-transmitted helminthiasis in Myanmar and approximate costs for countrywide control. Trop Med Int Health 9(9):1012–1015
Nakano T, Okamoto M, Ikeda Y, Hasegawa H (2006) Mitochondrial cytochrome *c* oxidase subunit 1 gene and nuclear rDNA regions of *Enterobius vermicularis* parasitic in captive chimpanzees with special reference to its relationship with pinworms in humans. Parasitol Res 100(1):51–57
Nithikathkul C, Changsap B, Wannapinyosheep S, Poister C, Boontan P (2001) The prevalence of *Enterobius vermicularis* among primary school students in Samut Prakan Province, Thailand. Southeast Asian J Trop Med Public Health 32(Suppl 2):133–137
Ok KS, Kim YS, Song JH, Lee JH, Ryu SH, Lee JH et al (2009) *Trichuris trichiura* infection diagnosed by colonoscopy: case reports and review of literature. Korean J Parasitol 47 (3):275–280
Pampiglione S, Ricciardi ML (1971) The presence of *Strongyloides fuelleborni* von Linstow, 1905, in man in Central and East Africa. Parassitologia 13(1):257–269
Panasoponkul C, Radomyos P, Singhasivanon V (1985) *Trichostrongylus* infection in a Thai boy. Southeast Asian J Trop Med Public Health 16(3):513–514
Phosuk I, Intapan PM, Sanpool O, Janwan P, Thanchomnang T, Sawanyawisuth K et al (2013a) Molecular evidence of *Trichostrongylus colubriformis* and *Trichostrongylus axei* infections in humans from Thailand and Lao PDR. Am J Trop Med Hyg 89(2):376–379
Phosuk I, Intapan PM, Thanchomnang T, Sanpool O, Janwan P, Laummaunwai P et al (2013b) Molecular detection of *Ancylostoma duodenale*, *Ancylostoma ceylanicum*, and *Necator americanus* in humans in northeastern and southern Thailand. Korean J Parasitol 51(6):747–749
Phosuk I, Sanpool O, Thanchomnang T, Sadaow L, Rodpai R, Anamnart W et al (2018) Molecular identification of *Trichuris suis* and *Trichuris trichiura* eggs in human populations from Thailand, Lao PDR, and Myanmar. Am J Trop Med Hyg 98(1):39–44
Polseela P, Poodendan W, Tangchaisuriya U, Nithikathkul C, Arnat N, Pannarunothai S et al (2004) Parasitic infection among primary school students in Mueang district, Phitsanulok Province, Thailand. Southeast Asian J Trop Med Public Health 35(Suppl 1):120–122
Pullan RL, Smith JL, Jasrasaria R, Brooker SJ (2014) Global numbers of infection and disease burden of soil transmitted helminth infections in 2010. Parasit Vectors 7:37
Repetto SA, Ruybal P, Batalla E, López C, Fridman V, Sierra M et al (2018) Strongyloidiasis outside endemic areas: long-term parasitological and clinical follow-up after ivermectin treatment. Clin Infect Dis 66(10):1558–1565

Requena-Méndez A, Chiodini P, Bisoffi Z, Buonfrate D, Gotuzzo E, Muñoz J (2013) The laboratory diagnosis and follow up of strongyloidiasis: a systematic review. PLoS Negl Trop Dis 7(1):e2002

Sadaow L, Sanpool O, Intapan PM, Sukeepaisarnjaroen W, Prasongdee TK, Maleewong W (2018a) A hospital-based study of intestinal capillariasis in Thailand: clinical features, potential clues for diagnosis, and epidemiological characteristics of 85 patients. Am J Trop Med Hyg 98(1):27–31

Sadaow L, Sanpool O, Phosuk I, Rodpai R, Thanchomnang T, Wijit A et al (2018b) Molecular identification of *Ascaris lumbricoides* and *Ascaris suum* recovered from humans and pigs in Thailand, Lao PDR, and Myanmar. Parasitol Res 117(8):2427–2436

Saichua P, Nithikathkul C, Kaewpitoon N (2008) Human intestinal capillariasis in Thailand. World J Gastroenterol 14(4):506–510

Saksirisampant W, Prownebon J, Kulkumthorn M, Yenthakam S, Janpla S, Nuchprayoon S (2006) Prevalence of intestinal parasitic infections among school children in the central region of Thailand. J Med Assoc Thai 89(11):1928–1933

Sandground JH (1925) Speciation and specificity in the nematode genus *Strongyloides*. J Parasitol 12(2):59–80

Sanpool O, Intapan PM, Rodpai R, Laoraksawong P, Sadaow L, Tourtip S et al (2019) Dogs are reservoir hosts for possible transmission of human strongyloidiasis in Thailand: molecular identification and genetic diversity of causative parasite species. J Helminthol 94:e110

Sato M, Sanguankiat S, Yoonuan T, Pongvongsa T, Keomoungkhoun M, Phimmayoi I et al (2010) Copro-molecular identification of infections with hookworm eggs in rural Lao PDR. Trans R Soc Trop Med Hyg 104(1):617–622

Sato M, Yoonuan T, Sanguankiat S, Nuamtanong S, Pongvongsa T, Phimmayoi I et al (2011) Short report: human *Trichostrongylus colubriformis* infection in a rural village in Laos. Am J Trop Med Hyg 84(1):52–54

Schär F, Trostdorf U, Giardina F, Khieu V, Muth S, Marti H et al (2013) *Strongyloides stercoralis*: global distribution and risk factors. PLoS Negl Trop Dis 7(7):e2288

Schneider B, Jariwala AR, Periago MV, Gazzinelli MF, Bose SN, Hotez PJ et al (2011) A history of hookworm vaccine development. Hum Vaccin 7(11):1234–1244

Senephansiri P, Laummaunwai P, Laymanivong S, Boonmar T (2017) Status and risk factors of *Strongyloides stercoralis* infection in rural communities of Xayaburi Province, Lao PDR. Korean J Parasitol 55(5):569–573

Shao CC, Xu MJ, Alasaad S, Song HQ, Peng L, Tao JP et al (2014) Comparative analysis of microRNA profiles between adult *Ascaris lumbricoides* and *Ascaris suum*. BMC Vet Res 10:99

Sharifdini M, Mirhendi H, Ashrafi K, Hosseini M, Mohebali M, Khodadadi H et al (2015) Comparison of nested polymerase chain reaction and real-time polymerase chain reaction with parasitological methods for detection of *Strongyloides stercoralis* in human fecal samples. Am J Trop Med Hyg 93(6):1285–1291

Sithithaworn P, Srisawangwong T, Tesana S, Daenseekaew W, Sithithaworn J, Fujimaki Y et al (2003) Epidemiology of *Strongyloides stercoralis* in north-east Thailand: application of the agar plate culture technique compared with the enzyme-linked immunosorbent assay. Trans R Soc Trop Med Hyg 97(4):398–402

Soukhathammavong P, Sayasone S, Harimanana AN, Akkhavong A, Thammasack S, Phoumindr N et al (2008) Three cases of intestinal capillariasis in Lao People's Democratic Republic. Am J Trop Med Hyg 79(5):735–738

Speare R (1989) Identification of species of *Strongyloides*. In: Grove DI (ed) Strongyloidiasis: a major roundworm infection of man. Taylor & Francis, London, pp 11–84

Stephenson LS, Holland CV, Cooper ES (2000) The public health significance of *Trichuris trichiura*. Parasitology 121(Suppl):S73–S95

Thanchomnang T, Intapan PM, Sanpool O, Rodpai R, Tourtip S, Yahom S et al (2017) First molecular identification and genetic diversity of *Strongyloides stercoralis* and *Strongyloides fuelleborni* in human communities having contact with long-tailed macaques in Thailand. Parasitol Res 116(7):1917–1923

Thanchomnang T, Intapan PM, Sanpool O, Rodpai R, Sadaow L, Phosuk I et al (2019) First molecular identification of *Strongyloides fuelleborni* in long-tailed macaques in Thailand and Lao People's Democratic Republic reveals considerable genetic diversity. J Helminthol 93 (5):608–615

Tomanakan K, Sanpool O, Chamavit P, Lulitanond V, Intapan PM, Maleewong W (2018) Genetic variation of *Enterobius vermicularis* among schoolchildren in Thailand. J Helminthol 94:e7

Traub RJ, Inpankaew T, Sutthikornchai C, Sukthana Y, Thompson RC (2008) PCR-based coprodiagnostic tools reveal dogs as reservoirs of zoonotic ancylostomiasis caused by *Ancylostoma ceylanicum* in temple communities in Bangkok. Vet Parasitol 155(1–2):67–73

Vince JD, Ashford RW, Gratten MJ, Bana-Koiri J (2005) *Strongyloides* species infestation in young infants of Papua New Guinea: association with generalized oedema. 1979. P N G Med J 48 (1–2):50–57

Wall EC, Bhatnagar N, Watson J, Doherty T (2011) An unusual case of hypereosinophilia and abdominal pain: an outbreak of *Trichostrongylus* imported from New Zealand. J Travel Med 18 (1):59–60

Wallace L, Henkin R, Mathies AW (1956) *Trichostrongylus* infestation with profound eosinophilia. Ann Intern Med 45(1):146–150

Wongsaroj T, Phatihatakorn W, Ramasoota P, Anamnart W, Kaewpoonsri N, Chiewchanyon B (2008) Epidemiological study of strongyloidiasis in southern Thailand. J Trop Med Parasitol 31:6–13

World Health Organization (1991) Basic laboratory methods in medical parasitology. 1. Parasitology manuals. https://www.who.int/malaria/publications/atoz/9241544104_part1/en/. Accessed 19 Jul 2020

World Health Organization (2019) Soil-transmitted helminth infections. https://www.who.int/news-room/fact-sheets/detail/soil-transmitted-helminth-infections. Accessed 27 Jun 2020

Yong TS, Lee JH, Sim S, Lee J, Min DY, Chai JY et al (2007) Differential diagnosis of *Trichostrongylus* and hookworm eggs via PCR using ITS-1 sequence. Korean J Parasitol 45(1):69–74

Yong TS, Chai JY, Sohn WM, Eom KS, Jeoung HG, Hoang EH et al (2014) Prevalence of intestinal helminths among inhabitants of Cambodia (2006–2011). Korean J Parasitol 52(6):661–666

Youssef FG, Mikhail EM, Mansour NS (1989) Intestinal capillariasis in Egypt: a case report. Am J Trop Med Hyg 40(2):195–196

# Chapter 6
# Molecular Identification and Genetic Diversity of Cestodes in Southeast Asia

**Tongjit Thanchomnang, Oranuch Sanpool, Hiroshi Yamasaki, Pewpan M. Intapan, and Wanchai Maleewong**

**Abstract** Southeast Asia is geographically divided into two sub-regions, Mainland Southeast Asia and Maritime Southeast Asia. The former includes Cambodia, Lao PDR, Myanmar (Burma), Peninsular Malaysia, Thailand, and Vietnam, and the latter includes Indonesia, the Philippines, East Malaysia, Brunei, Singapore, and East Timor. These tropical countries all have an environment and host populations suitable for infective helminths. Approximately ten million people in this region live in poor sanitary environments and are affected by a variety of neglected parasitic helminths and protozoan infections. In this chapter, we review the medically important tapeworms that are of public health concern in Southeast Asia with emphasis on molecular identification and genetic diversity.

**Keywords** *Taenia* spp. · *Echinococcus* spp. · *Hymenolepis* spp. · *Spirometra* spp. · *Sparganum proliferum* · *Dibothriocephalus* spp. · Molecular identification · Genetic diversity

---

T. Thanchomnang
Faculty of Medicine, Mahasarakham University, Maha Sarakham, Thailand
e-mail: tongjit.t@msu.ac.th

O. Sanpool · P. M. Intapan · W. Maleewong (✉)
Department of Parasitology, Faculty of Medicine, Khon Kaen University, Khon Kaen, Thailand

Mekong Health Science Research Institute, Khon Kaen University, Khon Kaen, Thailand
e-mail: oransa@kku.ac.th; pewpan@kku.ac.th; wanch_ma@kku.ac.th

H. Yamasaki
Department of Parasitology, National Institute of Infectious Diseases, Tokyo, Japan
e-mail: hyamasak@nih.go.jp



T. N. Petney et al. (eds.), *Biodiversity of Southeast Asian Parasites and Vectors causing Human Disease*, Parasitology Research Monographs 14,
https://doi.org/10.1007/978-3-030-71161-0_6

## 6.1 Introduction

Southeast Asia is geographically divided into two sub-regions, Mainland Southeast Asia and Maritime Southeast Asia, both of which are located in the sub-tropical and tropical zones. Approximately ten million people live in poor sanitary environments and economic poverty in this region and are exposed to diverse neglected parasitic helminths and protozoan infections, as well as other pathogens. In this chapter, we review the medically important tapeworms that affect public health in Southeast Asia.

Tapeworms causing human cestodiasis belong to the orders Cyclophyllidea and Diphyllobothriidea (Platyhelminthes: Cestoda), which show differences in morphology, life cycles, and pathology (Sato et al. 2015). The order Cyclophyllidea includes 15 families of which the Taeniidae and Hymenolepididae contain genera including human-infecting species; *Taenia* (Linnaeus, 1758), *Echinococcus* (Rudolphi, 1801), and *Hymenolepis* (Rudolphi, 1819), respectively (Jones et al. 1994). The genus *Taenia* consists of nearly 50 species (Loos-Frank 2000; Hoberg et al. 2001; Hoberg 2006), whereas the genus *Echinococcus* is composed of nine species (Nakao et al. 2007; Thompson 2017). Taeniosis, caused by *Taenia solium*, *Taenia saginata*, and *Taenia asiatica*, is an important disease with public health concerns in Southeast Asia (Ito et al. 2003a). Echinococcosis is not endemic in Southeast Asia, but sporadic cases have been reported (McManus 2010). Hymenolepidiosis in humans is caused by *Hymenolepis nana* and *Hymenolepis diminuta* that are also endemic in Southeast Asia.

The order Diphyllobothriidea (previously Pseudophyllidea) (Kuchta et al. 2008) contains a large group of tapeworms parasitic in mammals, birds, amphibians, and reptiles (Bray et al. 1994). The family Diphyllobothriidae includes the most important genera for human infections: *Spirometra* (Mueller, 1936), *Sparganum* (Diesing, 1854), *Dibothriocephalus* (Lühe, 1899), and *Diphyllobothrium* (Cobbold, 1858). The most important species in Southeast Asia are *Spirometra* spp., *Sparganum proliferum*, and *Dibothriocephalus* spp. (Chai et al. 2005; Scholz et al. 2009; Schauer et al. 2014; Kuchta et al. 2015a; Waeschenbach et al. 2017; Yamasaki 2018).

## 6.2 *Taenia saginata* (Beef Tapeworm), *Taenia solium* (Pork Tapeworm), *Taenia asiatica* (Asian Tapeworm)

### *6.2.1 Classification/Geographic Distribution/Biology and Infection*

The taxonomy of human-infecting species, *T. saginata, T. solium,* and *T. asiatica*, has been well documented (Ito et al. 2003a; Nakao et al. 2002; Jeon and Eom 2013; Flisser 2013; Sato et al. 2015). *Taenia solium* is phylogenetically different from

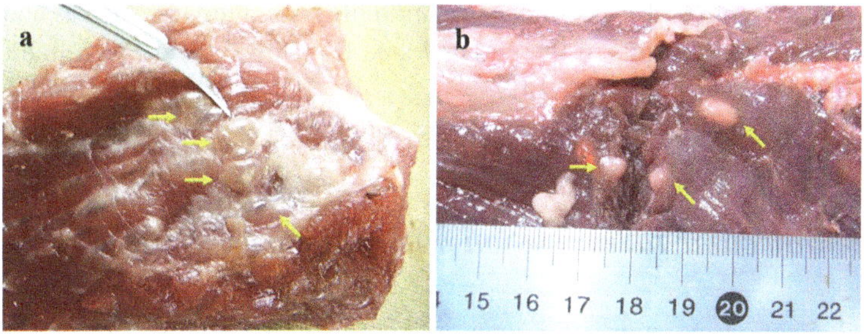


**Fig. 6.1** Taeniid cysticerci as sources of infection to humans. (**a**) *T. solium* cysticerci in pork, (**b**) *T. saginata* cysticerci in beef

*T. saginata* and *T. asiatica,* while *T. saginata* and *T. asiatica* are phylogenetically closely related and have diverged from a common ancestor (Hoberg et al. 2001; Hoberg 2006). However, natural hybrids between *T. asiatica* and *T. saginata* have been found in Thailand, Lao PDR, and China where these two species occur sympatrically, and taxonomic problems in the differentiation of *T. saginata* and *T. asiatica* have been raised (Okamoto et al. 2010; Yamane et al. 2012, 2013; Sato et al. 2018).

*Taenia solium* is distributed worldwide, except in some regions where people do not eat pork for religious reasons, and is endemic in Southeast Asian countries, Nepal and India (Schantz et al. 1998; Schantz 2006; Willingham 3rd and Engels 2006; Willingham 3rd et al. 2010; Ito et al. 2019). *Taenia saginata* also has a worldwide distribution and is endemic across Southeast Asia (Anantaphruti et al. 2013; Okamoto et al. 2010; Sanpool et al. 2017; Sato et al. 2018; Ito et al. 2019; Eichenberger et al. 2020). To date, *T. asiatica* is limited to 11 Asian countries (South Korea, Japan, China, Taiwan, the Philippines, Thailand, Vietnam, Lao PDR, Indonesia, Nepal and India) (Eom et al. 2009, 2020; Yamasaki et al. 2017).

Pigs serve as the intermediate hosts of *T. solium* and *T. asiatica* in which the oncosphere develops into a larval cysticercus. Cattle are the intermediate host for *T. saginata*. The larval cysticercus of *T. solium* develops in the muscle, tongue, and brain of pigs, but *T. asiatica* has a different viscerotropism from *T. solium*, occurring mainly in the pig liver (Eom et al. 2020). Humans are the only definitive hosts for these three *Taenia* species.

Humans become infected by eating raw or undercooked pork, pig viscera, or beef. The larva develops into an adult worm in the small intestine in humans and the adult tapeworms shed gravid proglottids filled with numerous eggs, which are shed with the stool. The eggs are ingested by intermediate hosts and develop into cysticerci in muscle tissue (Fig. 6.1) and other organs. In the case of *T. solium*, if humans ingest the eggs, oncospheres hatched from eggs can develop into cysticerci in subcutaneous tissues, muscle, and eyes, causing cysticercosis. The most serious form is neurocysticercosis that develops in the central nervous system (Schantz et al. 1998; Ito et al. 2003a, b). Neurocysticercosis is a major cause of epileptic seizures in endemic areas and is occasionally fatal (Schantz et al. 1998; Ito et al. 2003a, b). In contrast, *T. saginata* and *T. asiatica* cause mild illness in humans, such as abdominal pain, diarrhea, and pruritus in the perianal area (Pawlowski and Murrell 2001; Ooi et al. 2013; Yamasaki et al. 2021a).

### 6.2.2 Molecular Identification and Genetic Diversity

Morphological observation of human-infecting *Taenia* species is important for diagnosis (Fig. 6.2). However, classical identification is not always possible due to the morphological similarities of the eggs (Fig. 6.3), larval cysticerci and adult worms, especially in Asian regions where these three species occur sympatrically and human taeniosis is commonly reported as imported cases (Ito et al. 2003a, 2006; Yamasaki et al. 2004). The only reliable and accurate approaches for the

**Fig. 6.2** *Taenia saginata* adult worm recovered from an infected human stool in Lao PDR

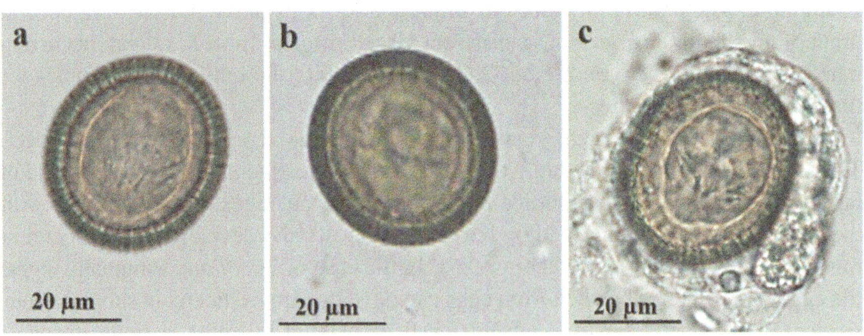


**Fig. 6.3** Human-*Taenia* eggs isolated from uteri in gravid proglottids. (**a**) *T. solium*, (**b**) *T. saginata*, (**c**) *T. asiatica*. Eggs are yellowish-brown in color, spherical in shape, and surrounded by embryophores with radial striations

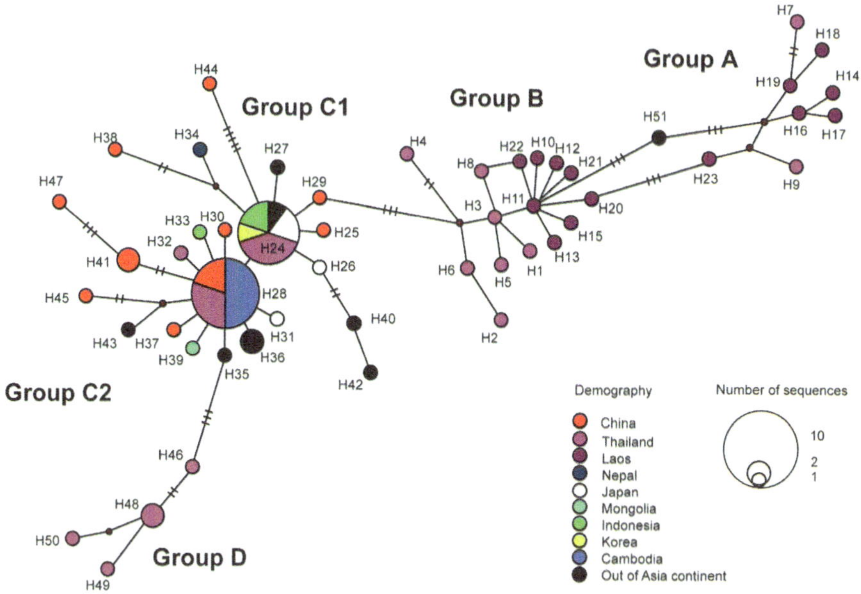


**Fig. 6.4** Median-joining haplotype network based on the complete *cox1* sequences of *Taenia saginata*. Each circle represents one haplotype and the size is proportional to the frequency of the haplotypes. Slashes indicate numbers of nucleotide differences inferred as having occurred between nodes. The colors represent the different countries

differentiation of the *Taenia* species are DNA-based methods. Several molecular methods targeting various DNA markers have been established: PCR-restriction fragment length polymorphism (RFLP) methods (Bowles and McManus 1994; Gasser and Chilton 1995), loop-mediated isothermal amplification (LAMP) (Nkouawa et al. 2012), species-specific DNA probes (Chapman et al. 1995), BESS T-base method (Yamasaki et al. 2002), PCR for the differentiation of *T. saginata* and *T. solium* (González et al. 2004), multiplex PCR (Yamasaki et al. 2004), a nested PCR system (Mayta et al. 2008), a high-resolution multiplex PCR assay (Jeon et al. 2009), and pyrosequencing (Thanchomnang et al. 2014). These methods are important not only for discriminating *Taenia* species but also for accumulating epidemiological and epizoological information (Yamasaki et al. 2004; Thanchomnang et al. 2014).

There have been several studies on the genetic diversity of human *Taenia* species (Nakao et al. 2002; Anantaphruti et al. 2013; Rostami et al. 2015; Solano et al. 2016; Sanpool et al. 2017). Anantaphruti et al. (2013) identified 14 haplotypes among 73 *T. saginata* samples from Thailand based on partial *cox1* sequences. Sanpool et al. (2017) identified 53 haplotypes among 98 complete *cox1* sequences in *T. saginata* from Asia and other continents. These haplotype network analyses demonstrate that *T. saginata* has a high genetic diversity and can be differentiated even in localities found close to one another within Southeast Asia (Anantaphruti et al. 2013; Sanpool et al. 2017). Haplotype network analyses revealed that global isolates of *T. saginata* are genetically divided into five groups (A, B, C1, C2 and D) (Fig. 6.4). Interestingly, *T. saginata* isolates from Lao PDR and northeastern Thailand belonged to either Group A or B. *Taenia saginata* from western Thailand clustered in groups C1, C2, and D, and populations from the northeast and western Thailand were found to be genetically distinct, suggesting that the spread of *T. saginata* in Thailand and Lao PDR is associated with different movement routes of human and cattle from China (Sanpool et al. 2017). Phylogenetic analysis of *cox1* sequences has revealed that *T. solium* is different from *T. saginata* and *T. asiatica,* while *T. saginata* and *T. asiatica* are closely related (Fig. 6.5).

## 6.3 *Echinococcus* Spp. (Hydatid Worm, Dog Tapeworm, Fox Tapeworm)

### *6.3.1 Classification/Geographic Distribution/Biology and Infection*

The taxonomy of the genus *Echinococcus* was controversial for many years; however, an accurate classification has been established based on molecular evidence (Nakao et al. 2007, 2013) and has been recently reviewed (Thompson 2017). The genus *Echinococcus* consists of 9 species: *E. granulosus, E. multilocularis, E. equinus, E. ortleppi, E. canadensis, E. felidis, E. shiquicus, E. vogeli* and *E. oligarthra*. *Echinococcus granulosus*, *E. ortleppi*, and *E. canadensis* have been reported as species causing cystic echinococcosis (CE) (Thompson 2017; Kern et al. 2017; Yousofi Darani and Jafari 2020), which is highly endemic in central Asia, western China, South America, eastern Africa, and Mediterranean countries (Wen et al. 2019). *Echinococcus multilocularis* is widely distributed in the northern hemisphere (from Eurasia to North America, including Japan), where it causes serious alveolar echinococcosis (AE) in humans (Wen et al. 2019; Deplazes et al. 2017; Vuitton et al. 2015; Gottstein et al. 2015). *Echinococcus felidis* and *E. shiquicus* are so far known from Africa and China, respectively. *Echinococcus vogeli* and *E. oligarthra* are limited to the Neotropics. CE and AE are not common in Southeast Asia (McManus 2010), however human cases have been sporadically reported in Thailand (Riengchan et al. 2004; Waikagul et al. 2006; Morakote et al. 2007) and Singapore (Teo et al. 1983). In Thailand, 22 cases of CE and two cases of AE were recorded from 1936 to 2005. Most Thai patients were migrant workers from the Middle East where CE is endemic, and only a few cases were indigenous (Waikagul et al. 2006), including a CE case reported in 2004 (Riengchan et al. 2004). A Thai patient with AE was recorded, but she was considered to have been infected in Switzerland where she lived from 1983 to 1993 (Warnnissorn et al. 2006). Two CE cases caused by infection with *E. ortleppi* were reported in Vietnam (De and Van 2017). Natural infections of *E. granulosus* in dogs were reported in

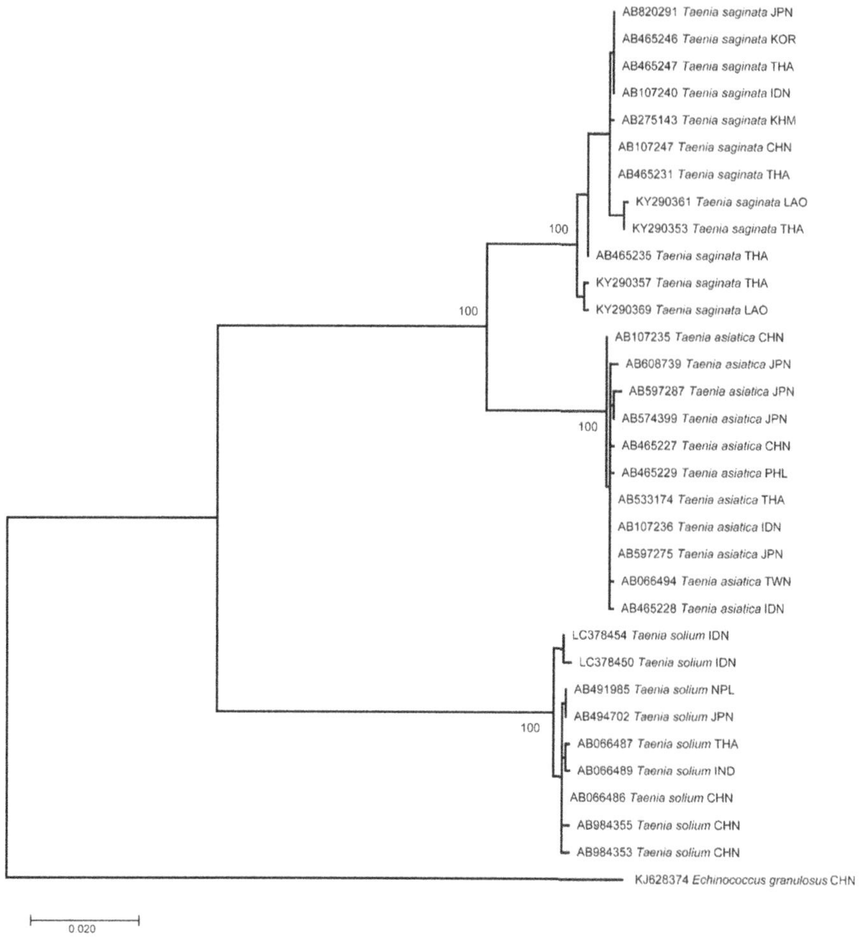


**Fig. 6.5** Maximum-likelihood phylogenetic tree based on the *cox*1 sequences (1620 bp) of *T. saginata, T. asiatica,* and *T. solium.* Bootstrap scores (>90%, percentages of 1000 pseudoreplications) are presented at each node. The sequences of *Taenia* species obtained from GenBank are indicated with their accession numbers and country codes (LAO, Lao People's Democratic Republic; THA, Thailand; IDN Indonesia; KHM, Cambodia; PHL, Philippines; IND, India; NPL, Nepal; JPN, Japan; KOR, Korea; TWN, Taiwan; CHN, China). *Echinococcus granulosus* was used as an outgroup

Sulawesi, Indonesia (Carney et al. 1974) and in wild dogs (*Cuon alpines* = *Cyon primaerus*) in South Vietnam (Le and Vu 1967).

Life cycles of *Echinococcus* species involve many herbivores and rodents as intermediate hosts and carnivores as the definitive hosts. The intermediate hosts become infected by ingesting the eggs, and the parasite then develops into larval stages in the visceral organs, mainly the liver. The canids, as definitive hosts, can be infected by ingesting the intermediate hosts as prey, and the larvae develop into adult tapeworms in the small intestine of these hosts. Humans act as accidental intermediate hosts and acquire an infection by ingesting eggs (Agudelo Higuita et al. 2016; Thompson 2017).

The diagnosis of AE in a Thai patient was based on the typical macroscopic multivesicular lesion and the characteristic histological findings of the parasite in the liver, the long-term stay of the patient in an endemic area, and evidence of an infection risk in the area. The rarity of AE in Thailand contributes to the lack of familiarity of the condition by medical personnel, the non-availability of diagnostic tools such as serologic tests, and inexperience in diagnosis and management (Warnnissorn et al. 2006). No echinococcosis diagnosed by DNA analysis has been reported for this area (McManus 2010).

## 6.4 *Hymenolepis* Spp. (Dwarf Tapeworm, Rat Tapeworm)

### *6.4.1 Classification/Geographic Distribution/Biology and Infection*

The family Hymenolepididae is a large group consisting of approximately 230 and 620 cestode species in mammals (rodents, insectivores, and bats, etc.) and birds, respectively (Haukisalmi et al. 2010). *Hymenolepis nana* and *Hymenolepis diminuta* are well-known human-infecting species, however the taxonomy of some species in the genus *Hymenolepis*, including *H. nana*, is still controversial (Haukisalmi et al. 2010).

*Hymenolepis nana* occurs worldwide and is one of the most common species among children under poor hygiene conditions, while *H. diminuta* has been less frequently found in various areas of the world, including Southeast Asia (Cross 1999). Human hymenolepidiosis caused by *H. nana* has been reported in Thailand (Sirivichayakul et al. 2000; Sithithaworn et al. 2003), Myanmar (Sahimin et al. 2016), Laos (Pakdeenarong et al. 2014; Ribas et al. 2017), Cambodia (Chhakda et al. 2006; Yong et al. 2014), Vietnam (Le Hung et al. 2005), and Indonesia (Cross et al. 1976; Toma et al. 1999; Prasetyo 2016). *Hymenolepis diminuta* infections in humans have also been reported in Thailand (Wiwanitkit 2004), Malaysia (Sinniah 1978; Rohela et al. 2012), and Indonesia (Cross et al. 1975; Stafford et al. 1980). *Hymenolepis nana, H. diminuta,* and *Hymenolepis* sp. have been found in rats in Malaysia and risks to human health are of concern (Paramasvaran et al. 2005; Zain et al. 2012).

Most of the hymenolepidid species require arthropod intermediate hosts, mainly small beetles (Coleoptera), in which eggs develop into cysticercoids. Adult worms parasitize the small intestine of the definitive hosts, rodents and humans (Cross 1999; Montgomery and Richards 2018). Human hymenolepidiosis due to *H. nana* is frequently contracted by human-to-human transmission through directly ingesting the eggs contaminating house dust, food, and water under poor hygienic conditions.

*Hymenolepis diminuta* infections are less frequent than *H. nana* infections, but occur in humans by the accidental ingestion of small beetles in stored cereal crops (Bogitsh et al. 2012; Montgomery and Richards 2018).

The molecular identification of *Hymenolepis* species has not been performed in Southeast Asian countries. Recently, molecular analysis has revealed that a species genetically closely related to *Hymenolepis hibernia* is a human-infecting species in China, and this species is genetically different from *H. nana* and *H. diminuta* (Nkouawa et al. 2016). Sargison et al. (2018) demonstrated that *H. diminuta* and *H. hibernia*, which are morphologically similar species, are genetically different. These findings indicate that humans might also be infected with unknown cryptic species. However, the identification of *H. diminuta* using molecular analysis is still rare in human infections in Southeast Asia. The molecular identification of hymenolepidid cestodes is necessary to determine the presence of cryptic species, to improve diagnosis, genetic characterization of the etiological agents, and epidemiological tracking of the disease (Mirdha and Rehman 2013).

## 6.5 *Spirometra* Spp. (Dog Tapeworm, Cat Tapeworm) and *Sparganum proliferum*

### *6.5.1 Classification/Geographic Distribution/Biology and Infection*

The genus *Spirometra* includes 46 nominal species (Scholz et al. 2019), of which only four are currently accepted as valid species (Kuchta and Scholz 2018; Scholz et al. 2019). However, the classification of the *Spirometra* species remains controversial and the validity of many species has not been verified (Waeschenbach et al. 2017; Kuchta and Scholz 2018; Scholz et al. 2019). *Spirometra erinaceieuropaei* is a well-known species causing human sparganosis and spirometriosis in Asia (Liu et al. 2015). Recently, *Spirometra decipiens* and *Spirometra ranarum* have been reported from Asia (Eom et al. 2015; Jeon et al. 2015, 2016, 2018a, b); however, the morphological identification of these three species, including *S. erinaceieuropaei* from Asia has been questioned (Waeschenbach et al. 2017; Kuchta and Scholz 2018; Scholz et al. 2019). Molecular revision of *Spirometra* species from Asia has revealed that two species occur in this area both of which are different from *S. erinaceieuropaei* from Europe (Yamasaki et al. 2021b). *Sparganum proliferum* is genetically classified in the family Diphyllobothriidae (Miyadera et al. 2001).

*Spirometra* species are widely distributed worldwide, including Southeast Asia (Lao PDR, Thailand, Myanmar, Cambodia, Vietnam, and Indonesia) (Jongthawin et al. 2014; Le et al. 2017). *Sparganum proliferum* has been found in Asia (South Korea, Japan, Taiwan, Thailand), Reunion Island (east of Madagascar), North and South Americas; however, the adult worm is unknown (Beaver et al. 1984; Miyazaki and Toh 1988; Miyazaki 1991; Kikuchi and Maruyama 2020).

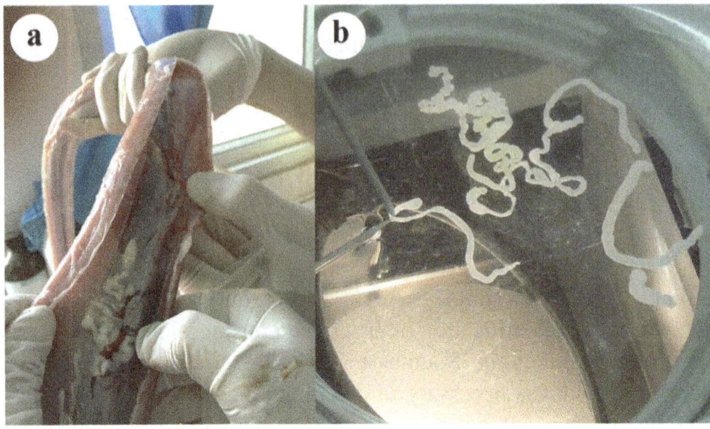


**Fig. 6.6** *Spirometra* plerocercoids (spargana). (**a**) Spargana in the subcutaneous tissues of Chinese rat snake (*Ptyas korros*) from Lao PDR, (**b**) Sparganum samples isolated from *P. korros* from Cambodia

*Spirometra* species require two intermediate hosts and the definitive hosts to complete their life cycles (Miyazaki and Toh 1988). The first intermediate hosts are freshwater copepods, in which a coracidium develops into a procercoid, and the second intermediate hosts in which a procercoid develops into a plerocercoid are amphibians and reptiles. Plerocercoids can survive for as long as a year in the second intermediate and paratenic hosts, such as amphibians, reptiles, birds, wild boar, and humans (Miyazaki and Toh 1988; Liu et al. 2010; Lee et al. 2013; Zhang et al. 2014; Jeon et al. 2016; Kołodziej-Sobocińska et al. 2016; Yamasaki et al. 2017) (Fig. 6.6). Canids and felids are the most common definitive hosts, in which the plerocercoid develops into an adult worm (Miyazaki and Toh 1988; Liu et al. 2015). *Sparganum proliferum* probably has a life cycle similar to other *Spirometra* species; however, nothing definite is known about this (Beaver et al. 1984; Miyazaki and Toh 1988; Miyazaki 1991). Humans are accidental hosts who become infected by ingesting the plerocercoids harbored in the flesh of the second intermediate or paratenic hosts and by drinking water polluted with cyclopoid copepods harboring the procercoids. The use of frog meat as a poultice is possibly a cause of the infection in certain Asian countries (Beaver et al. 1984; Miyazaki and Toh 1988; Miyazaki 1991).

Sparganosis is classified into two forms, non-proliferative and proliferative sparganosis. The former is caused by infection with *Spirometra* species, whereas the latter is caused by infection with *S. proliferum*. Non-proliferative sparganosis has been reported in Asia (Cho et al. 1975; Chung et al. 2000; Yoshikawa et al. 2010; Anantaphruti et al. 2011; Jongthawin et al. 2014; Kim et al. 2018; Li et al. 2019; Zhang et al. 2020), with the subcutaneous types being most common, especially in the thigh (Kudo et al. 2017), breast (Anantaphruti et al. 2011; Koonmee et al. 2011; Boonyasiri et al. 2013, 2014), abdominal subcutaneous tissues, and inguinal region (Yoshikawa et al. 2010), where the plerocercoid forms a nodule and presents as a creeping eruption. Ocular sparganosis is prevalent in Thailand, Vietnam, and China where people traditionally use frog-flesh poultices for open wounds and lesions of eyes (Miyazaki and Toh 1988; Wiwanitkit 2005; Anantaphruti et al. 2011). Plerocercoids can also invade the central nervous system and cause serious disease (Lv et al. 2010; Shirakawa et al. 2010; Boonyasiri et al. 2013, 2014) including blindness, paralysis, and even death. They are a major threat to human health (Qiu and Qiu 2009; Anantaphruti et al. 2011). In contrast, proliferative sparganosis is an extremely rare disease and 13 out of 18 cases reported from the world have been reported in Asia: six from Japan, three from Thailand, two from Taiwan, and one case each from South Korea and China (Kikuchi and Maruyama 2020). In cases of proliferative sparganosis the bones are destroyed by the invasion of worms. This has been diagnosed in two Thai patients (Settakorn et al. 2002; Laovachirasuwan et al. 2015), and two out of three fatal cases reported in Japan (Aoshima et al. 1989; Kikuchi and Maruyama 2020).

### *6.5.2 Molecular Identification and Genetic Diversity*

Adult worms of *Spirometra* species have been identified based on the morphology of their reproductive system, especially the number and shape of the uterine coils (Faust et al. 1929). However, since the morphology varies as tapeworms develop, it is not an appropriate criterion for identifying *Spirometra* species (Iwata 1972). Since definite morphological criteria for distinguishing *Spirometra* species have not been established, visual identifications can lead to many misidentifications and much confusion (Scholz et al. 2019). The most reliable tools for identifying plerocercoids and the adult worms of *Spirometra* species are molecular methods; however, only the DNA sequences of *S. proliferum* are available as credible reference data (Miyadera et al. 2001; Kikuchi et al. 2020).

The phylogenetic tree and haplotype network of *Spirometra* species from Asia based on complete *cox1* sequences are shown in Figs. 6.7 and 6.8, respectively. *Spirometra* species from Asia are divided into two clades, Type I and Type II, and haplotypes of Asian *Spirometra* Type I formed a phyletically differentiated network that is not star-like (Yamasaki et al. 2021b). It is interesting that *S. erinaceieuropaei*, *S. decipiens,* and *S. ranarum* from Asia are not monophyletic and are nested within Type I, indicating that Type I (= *Spirometra mansoni*) and Type II (= undescribed *Spirometra* sp.) are different species (Yamasaki et al. 2021b; Kuchta et al. 2021). The genetic diversity of Type I has been studied, but little is known about the genetic diversity of Type II because of the scarcity of samples examined (Yamasaki et al. 2021b).

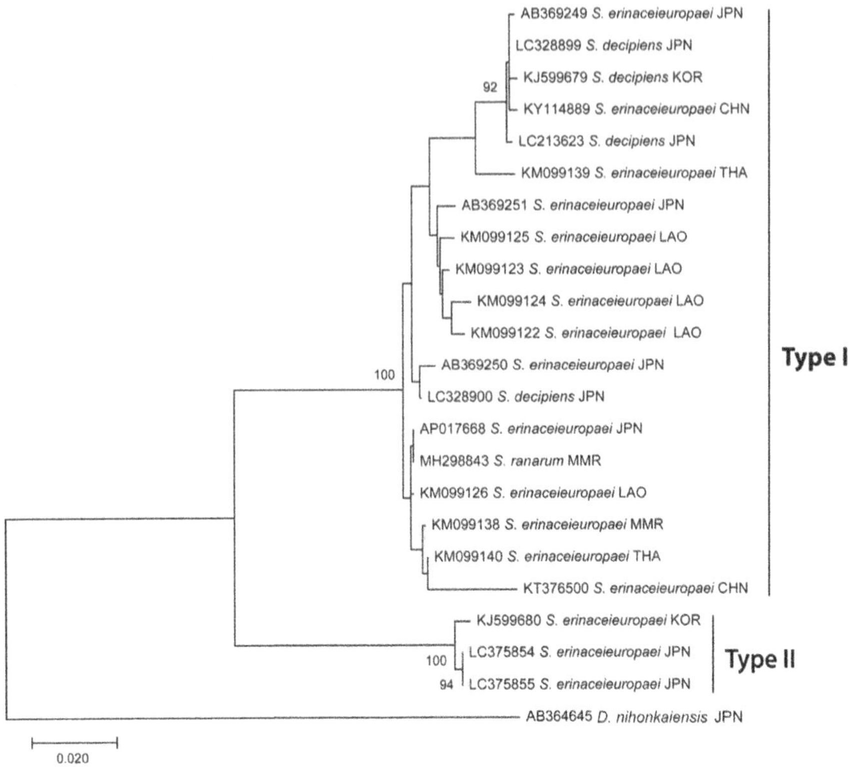


**Fig. 6.7** Maximum-likelihood phylogenetic tree based on the complete *cox1* sequences of *Spirometra* species from Asia. Bootstrap scores (>90%, percentages of 1000 pseudoreplications) are presented at each node. The sequences from GenBank are indicated with their accession numbers and country codes (LAO, Lao People's Democratic Republic; THA, Thailand; MMR, Myanmar; JPN, Japan; KOR, Korea; CHN, China). *Dibothriocephalus nihonkaiensis* was used as an outgroup

## 6.6 *Dibothriocephalus* and *Diphyllobothrium* Spp. (Broad Tapeworms)

### 6.6.1 *Classification/Geographic Distribution/Biology and Infection*

The genus *Diphyllobothrium* is currently assigned for broad tapeworms parasitic in marine mammals, such as whales and dolphins as the definitive hosts, while the genus *Dibothriocephalus* is assigned for broad tapeworms parasitic in terrestrial mammals and birds as the definitive hosts (Waeschenbach et al. 2017). The former includes *Diphyllobothrium stemmacephalum* (the type species of the genus) and *Diphyllobothrium balaenopterae* (syn. *Diplogonoporus balaenopterae*), and the latter includes well-known *Dibothriocephalus latus* (the type species of the genus,

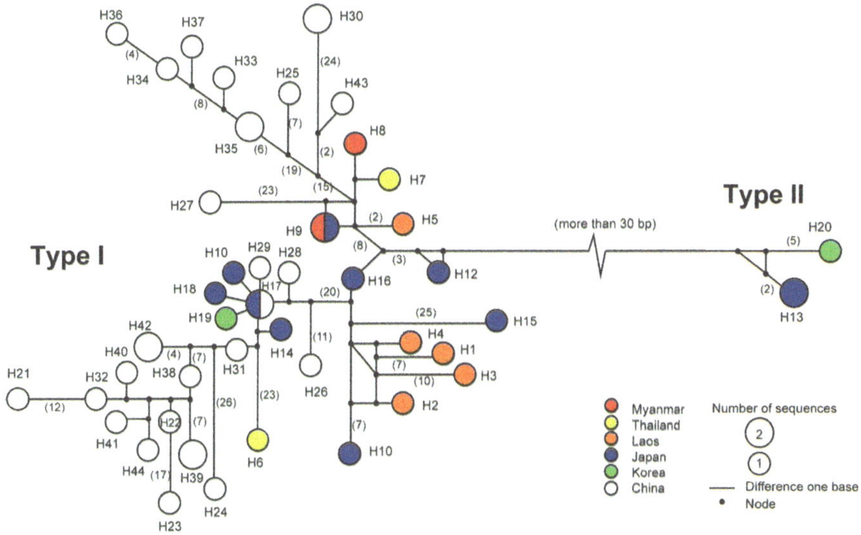


**Fig. 6.8** Median-joining haplotype network of *Spirometra* species from Asia based on the complete *cox1* sequences (1566 bp). Each circle represents one haplotype and the size of the circle is the number of the haplotype. Numbers in parentheses indicate numbers of nucleotide differences inferred as having occurred between nodes. The colors represent the different geographical origins (countries)

syn. *Diphyllobothrium latum*) and *Dibothriocephalus nihonkaiensis* (syn. *Diphyllobothrium nihonkaiense*). *Dibothriocephalus* species are distributed in the temperate to subarctic zones in Holarctic (Eurasia, North America), but only two species, *Dib. latus* and *Dibothriocephalus dendriticus,* occur in Neotropics (southern Chile and Argentine). *Dibothriocephalus latus* and *Dib. nihonkaiensis* are well known as the most frequent etiologic agents of human diphyllobothriosis (Dick et al. 2001; Chai et al. 2005; Scholz et al. 2009; Kuchta et al. 2015a; Yamasaki 2018). Dibothriocephalosis/diphyllobothriosis is an extremely rare disease in Southeast Asia. Two human cases due to *Dib. latus* infections had been reported in Malaysia in 2002 and 2006 (Rohela et al. 2002, 2006); however, the identification of the etiological agents was questionable (Yamasaki 2018). The increase of dibothriocephalosis due to *Dib. nihonkaiensis* is globally expected, including Southeast Asia, because of the globalization of the fresh or chilled Pacific salmon trade and the increasing popularity of raw salmon dishes, such as *sashimi* and *sushi* (Yamasaki 2018; Ikuno et al. 2018). Indeed, two human cases due to *D. nihonkaiensis*, diagnosed by *cox1* sequencing, were recently reported in Singapore with the possible source considered to be imported salmon (Ko et al. 2019).

Broad tapeworms require two intermediate hosts and the definitive host to complete their life cycles. The life cycle of *Dib. nihonkaiensis* is as follows (Scholz et al. 2009; Kuchta et al. 2015a; Yamasaki 2018): the first intermediate hosts are unknown, but are probably marine or blackish zooplanktonic copepods in which the coracidium develops into a procercoid. This is then ingested by the second intermediate hosts, Pacific salmon, in which the procercoid develops into a plerocercoid. The plerocercoid develops into an adult worm in the small intestine of the definitive hosts, including humans. The source of human infection is raw or undercooked Pacific salmon (Arizono et al. 2009). The adult worm matures within 1 month, and long strobilae are excreted from the anus while defecating. Despite being a long tapeworm, clinical signs are generally mild, and light abdominal pain and diarrhea may present occasionally. The diagnosis is routinely performed based on the morphology of the proglottids and eggs in the feces in the laboratory. *Dibothriocephalus* and *Diphyllobothrium* species are difficult to distinguish because of their morphological similarity. Dibothriocephalosis and diphyllobothriosis cases diagnosed by DNA-based methods are being increasingly reported. Thus, many molecular methods using different genetic markers have been developed for the identification of diphyllobothriid tapeworms: e.g. PCR and DNA sequencing (Yera et al. 2008; Wicht et al. 2010a), multiplex PCR (Wicht et al. 2010b), PCR-RFLP (Yamasaki et al. 2014), and pyrosequencing (Thanchomnang et al. 2016).

To date, no human infections caused by *Dip. stemmacephalum* and *Dip. balaenopterae* or *Adenocephalus pacificus* have been reported in Southeast Asia (Yamasaki et al. 2012, 2016; Kuchta et al. 2015a, b); however, it may be present in this region because the sources of infection to humans are marine fish that are a common food item throughout the area.

## 6.7 Conclusion Remarks and Research Needs

The molecular identification and genetic diversity of *Taenia* spp. and *Spirometra* spp., the common tapeworms causing human diseases in Southeast Asia, have been widely reported. However, the molecular study of *T.asiatica* is still limited and requires more consideration. Moreover, the larval agents causing cysticercosis remain a neglected topic in most part of the areas. The regional prevalence and molecular study of cyclophyllidean tapeworms, i.e. *Dipylidium caninum* and unclassified species, *Hymenolepis* spp. and *Echinococcus* spp. are still lacking. Furthermore, lack of knowledge of the interesting broad tapeworms belonging to the genera *Dibothriocephalus* and *Diphyllobothrium* requires additional research. It is notable that infections with these tapeworms are becoming important and globally due to the transport of fresh meat, the movement of people, changing consumption habits and food preparation, as well as personal hygiene. The incidences of the infections are possibly rising and need further studies. Presently, reliable molecular tools have been established for studying taxonomy and diversity. Our understanding of the molecular identification and genetic diversity of these parasites is necessary to help to improve diagnosis, genetic characterization of the etiological agents, epidemiological tracking and epizoology, as well as for the implementation of specific control measures. It is the responsibility of all stakeholders to force active networks linking researchers to define and implement surveillance and control methodologies.

**Acknowledgement** The authors (TT, OS, PMI and WM) were supported by a Distinguished Research Professor Grant, Thailand Research Fund (Grant no. DPG 6280002).

## References

Agudelo Higuita NI, Brunetti E, McCloskey C (2016) Cystic echinococcosis. J Clin Microbiol 54 (3):518–523

Anantaphruti MT, Nawa Y, Vanvanitchai Y (2011) Human sparganosis in Thailand: an overview. Acta Trop 118(3):171–176

Anantaphruti M, Thaenkham U, Kusolsuk T, Maipanich W, Saguankiat S, Pubampen S et al (2013) Genetic variation and population genetics of *Taenia saginata* in north and northeast Thailand in relation to *Taenia asiatica*. J Parasitol Res 2013:310605

Aoshima M, Nakata K, Matsuoka M, Kawabata M, Nakamura T (1989) A case of proliferative sparganosis associated with PIE syndrome and pulmonary embolism. Nihon Kyoubu-Shinkei-Gakkai Zasshi 27(12):1521–1527 (in Japanese with English summary)

Arizono N, Yamada M, Nakamura-Uchiyama F, Ohnishi K (2009) Diphyllobothriasis associated with eating raw pacific salmon. Emerg Infect Dis 15(6):866–870

Beaver PC, Jung RC, Cupp EW (1984) Pseudophyllidean tapeworms. In: Beaver PC, Jung RC, Cupp EW (eds) Clinical Parasitology, 9th edn. Lea & Febiger, Philadelphia, pp 494–504

Bogitsh BJ, Carter CE, Oeltmann TN (2012) Human parasitology. Academic Press, New York

Boonyasiri A, Cheunsuchon P, Srirabheebhat P, Yamasaki H, Maleewong W, Intapan PM (2013) Sparganosis presenting as cauda equina syndrome with molecular identification of the parasite in tissue sections. Korean J Parasitol 51(6):739–742

Boonyasiri A, Cheunsuchon P, Suputtamongkol Y, Yamasaki H, Sanpool O, Maleewong W et al (2014) Nine human sparganosis cases in Thailand with molecular identification of causative parasite species. Am J Trop Med Hyg 91(2):389–393

Bowles J, McManus DP (1994) Genetic characterization of the Asian *Taenia*, a newly described taeniid cestode of humans. Am J Trop Med Hyg 50(1):33–44

Bray RA, Jones A, Andersen KI (1994) Order Pseudophyllidea Carus, 1863. In: Khalil LF, Bray RA (eds) Keys to the cestode parasites of vertebrates. CAB International, Wallingford, pp 205–247

Carney WP, Cross JH, Wheeling CH, Purnomo SM, Simandjuntak G (1974) Natural infection of *Echinococcus granulosus* in dogs from Sulawesi, Indonesia. Southeast Asian J Trop Med Public Health 5(3):385–389

Chai JY, Murrell KD, Lymbery AJ (2005) Fish-borne parasitic zoonoses: status and issues. Int J Parasitol 35(11–12):1233–1254

Chapman A, Vallejo V, Mossie KG, Ortiz D, Agabian N, Flisser A (1995) Isolation and characterization of species-specific DNA probes from *Taenia solium* and *Taenia saginata* and their use in an egg detection assay. J Clin Microbiol 3(5):1283–1288

Chhakda T, Muth S, Socheat D, Odermatt P (2006) Intestinal parasites in school-aged children in villages bordering Tonle Sap Lake, Cambodia. Southeast Asian J Trop Med Public Health 37 (5):859–864

Cho SY, Bae JH, Seo BS (1975) Some aspects of human sparganosis in Korea. Kisaengchunghak Chapchi 13(1):60–77

Chung CC, Tsai BJ, Lin TY, Kuo HM (2000) Cutaneous sparganosis -a case report. Dermatol Sinica (Taiwan) 18:204–210 (English with Chinese abstract)

Cross JH (1999) Section 8 chapter 25 parasites: helminths. In: Berkley SF et al (eds) Infectious diseases. Harcourt Publishers, London, pp 8.35.1–8.35.18

Cross JH, Clarke MD, Durfee PT, Irving GS, Taylor J, Partono F et al (1975) Parasitology survey and seroepidemiology of amoebiasis in South Kalimantan (Borneo), Indonesia. Southeast Asian J Trop Med Public Health 6(1):52–60

Cross JH, Clarke MD, Cole WC, Lien JC, Partono F, Djakaria et al (1976) Parasitic infections in humans in West Kalimantan (Borneo), Indonesia. Trop Geogr Med 28(2):121–130

De NV, Van DL (2017) The first report of two cases of cystic echinococcosis in the lung by *Echinococcus ortleppi* infection, in Vietnam. Res Rep Trop Med 8:45–51

Deplazes P, Rinaldi L, Alvarez Rojas CA, Torgerson PR, Harandi MF, Romig T et al (2017) Global distribution of alveolar and cystic echinococcosis. Adv Parasitol 95:315–493

Dick TA, Nelson PA, Choudhury A (2001) Diphyllobothriasis: update on human cases, foci, patterns and sources of human infections and future considerations. Southeast Asian J Trop Med Public Health 32(Suppl 2):59–76

Eichenberger RM, Thomas LF, Gabriël S, Bobić B, Devleesschauwer B, Robertson LJ et al (2020) Epidemiology of *Taenia saginata* taeniosis/cysticercosis: a systematic review of the distribution in East, Southeast and South Asia. Parasit Vectors 13(1):234

Eom KS, Jeon HK, Rim HJ (2009) Geographical distribution of *Taenia asiatica* and related species. Korean J Parasitol 47(Suppl):S115–124

Eom KS, Park H, Lee D, Choe S, Kim KH, Jeon HK (2015) Mitochondrial genome sequences of *Spirometra erinaceieuropaei* and *S. decipiens* (Cestoidea: Diphyllobothriidae). Korean J Parasitol 53(4):455–463

Eom KS, Rim HJ, Jeon HK (2020) *Taenia asiatica*: historical overview of taeniasis and cysticercosis with molecular characterization. Adv Parasitol 108:133–173

Faust EC, Campbell HE, Kellogg CR (1929) Morphological and biological studies on the species of *Diphyllobothrium* in China. Am J Epidemiol 9:560–883

Flisser A (2013) State of the art of *Taenia solium* as compared to *Taenia asiatica*. Korean J Parasitol 51(1):43–49

Gasser RB, Chilton NB (1995) Characterisation of taeniid cestode species by PCR-RFLP of ITS2 ribosomal DNA. Acta Trop 59(1):31–40

González LM, Montero E, Morakote N, Puente S, Díaz De Tuesta JL, Serra T et al (2004) Differential diagnosis of *Taenia saginata* and *Taenia saginata asiatica* taeniasis through PCR. Diagn Microbiol Infect Dis 49(3):183–188

Gottstein B, Stojkovic M, Vuitton DA, Millon L, Marcinkute A, Deplazes P (2015) Threat of alveolar echinococcosis to public health--a challenge for Europe. Trends Parasitol 31 (9):407–412

Haukisalmi V, Hardman LM, Foronda P, Feliu C, Laakkonen J, Niemimaa J et al (2010) Systematic relationships of hymenolepidid cestodes of rodents and shrews inferred from sequences of 28S ribosomal RNA. Zool Scr 39:631–641

Hoberg EP (2006) Phylogeny of *Taenia*: species definitions and origins of human parasites. Parasitol Int 55(Suppl):S23–S30

Hoberg EP, Alkire N, de Quieroz A, Jones A (2001) Out of Africa: origins of the *Taenia* tapeworms in humans. Proc Bio Sci 268(1469):781–787

Ikuno H, Akao S, Yamasaki H (2018) Epidemiology of *Diphyllobothrium nihonkaiense* diphyllobothriasis, Japan, 2001-2016. Emerg Infect Dis 24(8):1428–1434

Ito A, Nakao M, Wandra T (2003a) Human taeniasis and cysticercosis in Asia. Lancet 362 (9399):1918–1920

Ito A, Yamasaki H, Nakao M, Sako Y, Okamoto M, Sato MO et al (2003b) Multiple genotypes of *Taenia solium*--ramifications for diagnosis, treatment and control. Acta Trop 87(1):95–101

Ito A, Takayanagui OM, Sako Y, Sato MO, Odashima NS, Yamasaki H et al (2006) Neurocysticercosis: clinical manifestation, neuroimaging, serology and molecular confirmation of histopathologic specimens. Southeast Asian J Trop Med Public Health 37(Suppl 3):74–81

Ito A, Li T, Wandra T, Dekumyoy P, Yanagida T, Okamoto M et al (2019) Taeniasis and cysticercosis in Asia: a review with emphasis on molecular approaches and local lifestyles. Acta Trop 198:105075

Iwata S (1972) Experimental and morphological studies of Manson's tapeworm *Diphyllobothrium erinacei* (Rudolphi) special reference to its scientific name and relationship with *Sparganum proliferum* Ijima. Prog Med Parasitol Jpn 4:536–590 (Meguro Parasitological Museum, Tokyo)
Jeon HK, Eom KS (2013) Molecular approaches to *Taenia asiatica*. Korean J Parasitol 51(1):1–8
Jeon HK, Chai JY, Kong Y, Waikagul J, Insisiengmay B, Rim HJ et al (2009) Differential diagnosis of *Taenia asiatica* using multiplex PCR. Exp Parasitol 121(2):151–156
Jeon HK, Park H, Lee D, Choe S, Kim KH, Huh S et al (2015) Human infections with *Spirometra decipiens* plerocercoids identified by morphologic and genetic analyses in Korea. Korean J Parasitol 53(3):299–305
Jeon HK, Park H, Lee D, Choe S, Kim KH, Sohn WM et al (2016) Genetic identification of *Spirometra decipiens* plerocercoids in terrestrial snakes from Korea and China. Korean J Parasitol 54(2):181–185
Jeon HK, Park H, Lee D, Choe S, Kang Y, Bia MM et al (2018a) Genetic and morphologic identification of *Spirometra ranarum* in Myanmar. Korean J Parasitol 56(3):275–280
Jeon HK, Huh S, Sohn WM, Chai JY, Eom KS (2018b) Molecular genetic findings of *Spirometra decipiens* and *S. ranarum* in Korea. Korean J Parasitol 56(4):359–364
Jones A, Bray RA, Khalil LF (1994) Order Cyclophyllidea van Beneden in Braun, 1900. In: Khalil LF, Jones A, Bray RA (eds) Key to the cestode parasites of vertebrates. CAB International, Wallingford, p 305
Jongthawin J, Intapan PM, Sanpool O, Sadaow L, Laymanivong S, Thanchomnang T et al (2014) Molecular evidence of *Spirometra erinaceieuropaei* infection in snakes *Ptyas korros* from LAO PDR and Thailand and frogs *Hoplobatrachus rugulosus* from Myanmar. Southeast Asian J Trop Med Public Health 45(6):1271–1278
Kern P, Menezes da Silva A, Akhan O, Müllhaupt B, Vizcaychipi KA, Budke C et al (2017) The echinococcosis: diagnosis, clinical management and burden of disease. Adv Parasitol 96:259–369
Kikuchi T, Maruyama H (2020) Human proliferative sparganosis update. Parasitol Int 75:102036
Kikuchi T, Dayi M, Hunt VL, Toyoda A, Maeda Y, Kondo Y et al (2020) Genome analysis of the fatal tapeworm *Sparganum proliferum* unravels the cryptic lifecycle and mechanisms underlying the aberrant larval proliferation. https://doi.org/10.1101/2020.05.19.10538
Kim JG, Ahn CS, Sohn WM, Nawa Y, Kong Y (2018) Human sparganosis in Korea. J Korean Med Sci 33(44):e273
Ko KKK, Cao DYH, Wee LWY, Hsu LY, Yamasaki H (2019) First two confirmed cases of *Dibothriocephalus nihonkaiensis* in Southeast Asia (Singapore). Pathology 51(4):433–434
Kołodziej-Sobocińska M, Miniuk M, Ruczyńska I, Tokarska M (2016) Sparganosis in wild boar (*Sus scrofa*) - implications for veterinarians, hunters, and consumers. Vet Parasitol 227:115–117
Koonmee S, Intapan PM, Yamasaki H, Sugiyama H, Muto M, Kuramochi T et al (2011) Molecular identification of a causative parasite species using formalin-fixed paraffin embedded (FFPE) tissues of a complicated human pulmonary sparganosis case without decisive clinical diagnosis. Parasitol Int 60(4):460–464
Kuchta R, Scholz T (2018) Diphyllobothriidea. In: Caira JN, Jensen K, Barbeau E (eds) Planetary biodiversity inventory (2008–2017): tapeworms from vertebrate bowels of the earth. University of Kansas, Natural History Museum, Special Publication No. 25, Lawrence, KS, pp 176–189
Kuchta R, Scholz T, Brabec J, Bray RA (2008) Suppression of the tapeworm order Pseudophyllidea (Platyhelminthes: Eucestoda) and the proposal of two new orders, Bothriocephalidea and Diphyllobothriidea. Int J Parasitol 38(1):49–55
Kuchta R, Scholz T, Brabec J, Wicht B (2015a) *Diphyllobothrium*, *Diplogonoporus*, and *Spirometra*. In: Xiao L, Ryan U, Feng Y (eds) Biology of foodborne parasites. CRC Press, Taylor & Francis Group, Boca Raton, FL, pp 299–326
Kuchta R, Serrano-Martínez ME, Scholtz T (2015b) Pacific broad tapeworm *Adenocephalus pacificus* as a causative agent of globally reemerging diphyllobothriosis. Emerg Infect Dis 21 (10):1697–1703

Kuchta R, Kołodziej-Sobocińska, Brabec J, Młocicki D, Sałamatin R, Scholz T (2021) Sparganosis (*Spirometra*) in Europe in the molecular era. Clin Infect Dis 72(5):882–890
Kudo T, Fujioka A, Korenaga M, Yamasaki H, Morishima Y, Sugiyama H et al (2017) Molecular identification of intramuscular and subcutaneous *Spirometra erinaceieuropaei* sparganosis in a Japanese patient. J Dermatol 44(6):e138–e139
Laovachirasuwan P, Mootsikapun P, Maleewong W, Intapan PM (2015) Intraosseous proliferative sparganosis presenting as a pathological fracture: a case report and review of the literature. Southeast Asian J Trop Med Public Health 46(4):558–563
Le Hung Q, de Vries PJ, Giao PT, Binh TQ, Nam NV, Kager PA (2005) Intestinal helminth infection in an ethnic minority commune in southern Vietnam. Southeast Asian J Trop Med Public Health 36(3):623–628
Le VH, Vu NT (1967) Concerning the presence of cestodes, *Echinococcus granulosus* (Batsch, 1786), in a wild dog, *Cyon primaerus* (Hodgs) in South Vietnam. Bull Soc Pathol Exot Filiales 60(1):64–71
Le AT, Do LT, Nguyen HT, Nguyen HT, Do AN (2017) Case report: the first case of human infection by adult of *Spirometra erinaceieuropaei* in Vietnam. BMC Infect Dis 17(1):669
Lee SH, Choe EY, Shin HD, Seo M (2013) Spargana in a weasel, *Mustela sibirica manchurica*, and wild boar, *Sus scrofa*, from Gangwon-do, Korea. Korean J Parasitol 51(3):379–381
Li H, Hu J, Yang P (2019) Diagnosis and treatment of human sparganosis. Lancet Infect Dis 19 (6):577–578
Liu W, Zhao GH, Tan MY, Zeng DL, Wang KZ, Yuan ZG et al (2010) Survey of *Spirometra erinaceieuropaei* spargana infection in the frog *Rana nigromaculata* of the Hunan Province of China. Vet Parasitol 173(1-2):152–156
Liu Q, Li MW, Wang ZD, Zhao GH, Zhu XQ (2015) Human sparganosis, a neglected food borne zoonosis. Lancet Infect Dis 15(10):1226–1235
Loos-Frank B (2000) An up-date of Verster's (1969) 'Taxonomic revision of the genus *Taenia* Linnaeus' (Cestoda) in table format. Syst Parasitol 45(3):155–183
Lv S, Zhang Y, Steinmann P, Zhou XN, Utzinger J (2010) Helminth infections of the central nervous system occurring in Southeast Asia and the Far East. Adv Parasitol 72:351–408
Mayta H, Gilman RH, Prendergast E, Castillo JP, Tinoco YO, Garcia HH et al (2008) Nested PCR for specific diagnosis of *Taenia solium* taeniasis. J Clin Microbiol 46(1):286–289
McManus DP (2010) Echinococcosis with particular reference to Southeast Asia. Adv Parasitol 72:267–303
Mirdha BR, Rehman S (2013) Hymenolepis. In: Liu D (ed) Molecular detection of human parasitic pathogens. CRC Press, Boca Raton, FL, pp 265–275
Miyadera H, Kokaze A, Kuramochi T, Kita K, Machinami R, Noya O et al (2001) Phylogenetic identification of *Sparganum proliferum* as a pseudophyllidean cestode by the sequence analyses on mitochondrial COI and nuclear sdhB genes. Parasitol Int 50(2):93–104
Miyazaki I (1991) Spirometriasis. In: Miyazaki I (ed) An illustrated book of helminthic zoonoses. International Medical Foundation of Japan, Tokyo, pp 207–214
Miyazaki I, Toh Y (1988) Parasitic Zoonoses. Kyushu University Press, Fukuoka (in Japanese)
Montgomery SP, Richards FO (2018) *Diphyllobothrium*, *Dipylidium*, and *Hymenolepis* species. In: Long SS, Prober CG, Fisher M (eds) Principles and practice of pediatric infectious diseases. Elsevier, Philadelphia, pp 1394–1397
Morakote N, Thamprasert K, Lojanapiwat B, Muttarak M (2007) Cystic echinococcosis in Thailand with a special note on detection by serology in one family. Southeast Asian J Trop Med Public Health 38(5):796–798
Nakao M, Okamoto M, Sako Y, Yamasaki H, Nakaya K, Ito A (2002) A phylogenetic hypothesis for the distribution of two genotypes of the pig tapeworm *Taenia solium* worldwide. Parasitology 124(6):657–662
Nakao M, McManus DP, Schantz PM, Craig PS, Ito A (2007) A molecular phylogeny of the genus *Echinococcus* inferred from complete mitochondrial genomes. Parasitology 134(5):713–722

Nakao M, Lavikainen A, Yanagida T, Ito A (2013) Phylogenetic systematics of the genus *Echinococcus* (Cestoda: Taeniidae). Int J Parasitol 43(12–13):1017–1029

Nkouawa A, Sako Y, Li T, Chen X, Nakao M, Yanagida T et al (2012) A loop-mediated isothermal amplification method for a differential identification of *Taenia* tapeworms from human: application to a field survey. Parasitol Int 61(4):723–725

Nkouawa A, Haukisalmi V, Li TY, Nakao M, Lavikainen A, Chen XW et al (2016) Cryptic diversity in hymenolepidid tapeworms infecting humans. Parasitol Int 65(2):83–86

Okamoto M, Nakao M, Blair D, Anantaphruti MT, Waikagul J, Ito A (2010) Evidence of hybridization between *Taenia saginata* and *Taenia asiatica*. Parasitol Int 59(1):70–74

Ooi HK, Ho CM, Chung WC (2013) Historical overview of *Taenia asiatica* in Taiwan. Korean J Parasitol 51(1):31–36

Pakdeenarong N, Siribat P, Chaisiri K, Douangboupha B, Ribas A, Chaval Y et al (2014) Helminth communities in murid rodents from southern and northern localities in Lao PDR: the role of habitat and season. J Helminthol 88(3):302–329

Paramasvaran S, Krishnasamy M, Lee HL, John J, Lokman H, Naseem BM et al (2005) Helminth infections in small mammals from Ulu Gombak Forest Reserve and the risk to human health. Trop Biomed 22(2):191–194

Pawlowski ZS, Murrell KD (2001) Taeniasis and cysticercosis. In: Hui YH, Sattar SA, Murrell KD, Nip W-K, Stanfield PS (eds) Foodborne disease handbook, 2nd edn. Marcel Dekker, New York, pp 217–227

Prasetyo RH (2016) Survey of house rat intestinal parasites from Surabaya district, east Java, Indonesia that can cause opportunistic infection in humans. Southeast Asian J Trop Med Public Health 47(2):194–198

Qiu MH, Qiu MD (2009) Human plerocercoidosis and sparganosis: II. A historical review on pathology, clinics, epidemiology and control. Chin J Parasitol Parasit Dis 27(3):251–260 (in Chinese with English abstract)

Ribas A, Jollivet C, Morand S, Thongmalayvong B, Somphavong S, Siew CC et al (2017) Intestinal parasitic infections and environmental water contamination in a rural village of northern Lao PDR. Korean J Parasitol 55(5):523–532

Riengchan P, Suankratay C, Wilde H, Thanakit V (2004) Hydatid disease of the liver: the first indigenous case in Thailand and review of the literature. J Med Assoc Thai 87(6):725–729

Rohela M, Jamaiah I, Chan KW, Yusoff WS (2002) Diphyllobothriasis: the first case report from Malaysia. Southeast Asian J Trop Med Public Health 33(2):229–230

Rohela M, Jamaiah I, Goh KL, Nissapatorn V (2006) A second case of diphyllobothriasis in Malaysia. Southeast Asian J Trop Med Public Health 37(5):896–898

Rohela M, Ngui R, Lim YA, Kalaichelvan B, Wan Hafiz WI, Mohd Redzuan AN (2012) A case report of *Hymenolepis diminuta* infection in a Malaysian child. Trop Biomed 29(2):224–230

Rostami S, Salavati R, Beech RN, Babaei Z, Sharbatkhori M, Harandi MF (2015) Genetic variability of *Taenia saginata* inferred from mitochondrial DNA sequences. Parasitol Res 114 (4):1365–1376

Sahimin N, Lim YA, Ariffin F, Behnke JM, Lewis JW, Mohd Zain SN (2016) Migrant workers in Malaysia: current implications of sociodemographic and environmental characteristics in the transmission of intestinal parasitic infections. PLoS Negl Trop Dis 10(11):e0005110

Sanpool O, Rodpai R, Intapan PM, Sadaow L, Thanchomnang T, Laymanivong S et al (2017) Genetic diversity of *Taenia saginata* (Cestoda: Cyclophyllidea) from Lao People's Democratic Republic and northeastern Thailand based on mitochondrial DNA. Parasit Vectors 10(1):141

Sargison N, Herman J, Pilkington J, Buckland P, Watt K, Chambers A, Chaudhry U (2018) Molecular confirmation of *Hymenolepis hibernia* in field mice (*Apodemus sylvaticus*) from St Kilda has potential to resolve a host-parasite relationship. Int J Parasitol Parasites Wildl 7 (3):364–368

Sato MO, Nunes CM, Sato M, Waikagul J (2015) Taenia. In: Xiao L, Ryan U, Feng Y (eds) Biology of foodborne parasites. CRC press, Taylor & Francis Group, New York, pp 463–480

Sato MO, Sato M, Yanagida T, Waikagul J, Pongvongsa T, Sako Y et al (2018) *Taenia solium*, *Taenia saginata*, *Taenia asiatica*, their hybrids and other helminthic infections occurring in a neglected tropical diseases' highly endemic area in Lao PDR. PLoS Negl Trop Dis 12(2): e0006260

Schantz PM (2006) Progress in diagnosis, treatment and elimination of echinococcosis and cysticercosis. Parasitol Int 55(Suppl):S7–S13

Schantz PM, Wilkins PP, Tsang VCW (1998) Immigrants, imaging and immunoblots: the emergence of neurocysticercosis as a significant public health problem. In: Scheld WM, Craig WA, Hughes JM (eds) Emerging infections. ASM Press, Washington, DC, p 213

Schauer F, Poppert S, Technau-Hafsi K, Mockenhaupt M, Muntau B, Häcker G et al (2014) Travel-acquired subcutaneous *Sparganum proliferum* infection diagnosed by molecular methods. Br J Dermatol 170(3):741–744

Scholz T, Garcia HH, Kuchta R, Wicht B (2009) Update on the human broad tapeworm (genus *Diphyllobothrium*), including clinical relevance. Clin Microbiol Rev 22(1):146–160

Scholz T, Kuchta R, Brabec J (2019) Broad tapeworms (Diphyllobothriidae), parasites of wildlife and humans: recent progress and future challenges. Int J Parasitol Parasites Wildl 9:359–369

Settakorn J, Arpornchayanon O, Chaiwun B, Vanittanakom P, Thamprasert K, Rangdaeng S (2002) Intraosseous proliferative sparganosis: a case report and review of the literature. J Med Assoc Thai 85(1):107–113

Shirakawa K, Yamasaki H, Ito A, Miyajima H (2010) Cerebral sparganosis: the wandering lesion. Neurology 74(2):180

Sinniah B (1978) *Hymenolepis diminuta* infection in a Malaysian oil palm estate worker -- first case from Malaysia. Southeast Asian J Trop Med Public Health 9(3):453–454

Sirivichayakul C, Radomyos P, Praevanit R, Pojjaroen-Anant C, Wisetsing P (2000) *Hymenolepis nana* infection in Thai children. J Med Assoc Thai 83(9):1035–1038

Sithithaworn P, Srisawangwong T, Tesana S, Daenseekaew W, Sithithaworn J, Fujimaki Y et al (2003) Epidemiology of *Strongyloides stercoralis* in north-east Thailand: application of the agar plate culture technique compared with the enzyme-linked immunosorbent assay. Trans R Soc Trop Med Hyg 97(4):398–402

Solano D, Navarro JC, León-Reyes A, Benítez-Ortiz W, Rodríguez-Hidalgo R (2016) Molecular analyses reveal two geographic and genetic lineages for tapeworms, *Taenia solium* and *Taenia saginata*, from Ecuador using mitochondrial DNA. Exp Parasitol 171:49–56

Stafford EE, Sudomo M, Masri S, Brown RJ (1980) Human parasitoses in Bali, Indonesia. Southeast Asian J Trop Med Public Health 11(3):319–323

Teo SK, Boey ML, Tan TH, Sng I (1983) Pulmonary hydatid cyst in Singapore--a case report. Singapore Med J 24(1):48–55

Thanchomnang T, Tantrawatpan C, Intapan PM, Sanpool O, Janwan P, Lulitanond V et al (2014) Rapid molecular identification of human taeniid cestodes by pyrosequencing approach. PLoS One 9(6):e100611

Thanchomnang T, Tantrawatpan C, Intapan PM, Sanpool O, Lulitanond V, Tourtip S et al (2016) Rapid identification of nine species of diphyllobothriidean tapeworms by pyrosequencing. Sci Rep 6:37228

Thompson RC (2017) Biology and systematics of *Echinococcus*. Adv Parasitol 95:65–109

Toma A, Miyagi I, Kamimura K, Tokuyama Y, Hasegawa H, Selomo M et al (1999) Questionnaire survey and prevalence of intestinal helminthic infections in Barru, Sulawesi, Indonesia. Southeast Asian J Trop Med Public Health 30(1):68–77

Vuitton DA, Demonmerot F, Knapp J, Richou C, Grenouillet F, Chauchet A et al (2015) Clinical epidemiology of human AE in Europe. Vet Parasitol 213(3–4):110–120

Waeschenbach A, Brabec J, Scholz T, Littlewood DTJ, Kuchta R (2017) The catholic taste of broad tapeworms - multiple routes to human infection. Int J Parasitol 47(13):831–843

Waikagul J, Dekumyoy P, Anantaphruti MT (2006) Taeniasis, cysticercosis and echinococcosis in Thailand. Parasitol Int 55(Suppl):S175–S180

Warnnissorn N, Uiprasertkul M, Atisook K, Sirivatanauksorn Y, Limawongpranee S (2006) Alveolar echinococcosis in a Thai patient after migration to an endemic area in Central Europe. Travel Med Infect Dis 4(1):34–37

Wen H, Vuitton L, Tuxun T, Li J, Vuitton DA, Zhang W, McManus DP (2019) Echinococcosis: advances in the 21st century. Clin Microbiol Rev 32(2):e00075–e00018

Wicht B, Ruggeri-Bernardi N, Yanagida T, Nakao M, Peduzzi R, Ito A (2010a) Inter- and intra-specific characterization of tapeworms of the genus *Diphyllobothrium* (Cestoda: Diphyllobothriidea) from Switzerland, using nuclear and mitochondrial DNA targets. Parasitol Int 59(1):35–39

Wicht B, Yanagida T, Scholz T, Ito A, Jiménez JA, Brabec J (2010b) Multiplex PCR for differential identification of broad tapeworms (Cestoda: *Diphyllobothrium*) infecting humans. J Clin Microbiol 48(9):3111–3116

Willingham AL 3rd, Engels D (2006) Control of *Taenia solium* cysticercosis/taeniosis. Adv Parasitol 61:509–566

Willingham AL 3rd, Wu HW, Conlan J, Satrija F (2010) Combating *Taenia solium* cysticercosis in Southeast Asia an opportunity for improving human health and livestock production. Adv Parasitol 72:235–266

Wiwanitkit V (2004) Overview of *Hymenolepis diminuta* infection among Thai patients. Med Gen Med 6(2):7

Wiwanitkit V (2005) A review of human sparganosis in Thailand. Int J Infect Dis 9(6):312–316

Yamane K, Suzuki Y, Tachi E, Li T, Chen X, Nakao M et al (2012) Recent hybridization between *Taenia asiatica* and *Taenia saginata*. Parasitol Int 61(2):351–355

Yamane K, Yanagida T, Li T, Chen X, Dekumyoy P, Waikagul J et al (2013) Genotypic relationships between *Taenia saginata*, *Taenia asiatica* and their hybrids. Parasitology 140 (13):1595–1601

Yamasaki H (2018) *Diphyllobothrium*, *Adenocephalus*, and *Diplogonoporus*. In: Liu D (ed) Food microbiology series: handbook of foodborne diseases. CRC Press. Taylor& Francis Group, Boca Raton, FL, pp 667–687

Yamasaki H, Nakao M, Sako Y, Nakaya K, Sato MO, Mamuti W et al (2002) DNA differential diagnosis of human taeniid cestodes by base excision sequence scanning thymine-base reader analysis with mitochondrial genes. J Clin Microbiol 40(10):3818–3821

Yamasaki H, Allan JC, Sato MO, Nakao M, Sako Y, Nakaya K et al (2004) DNA differential diagnosis of taeniasis and cysticercosis by multiplex PCR. J Clin Microbiol 42(2):548–553

Yamasaki H, Ohmae H, Kuramochi T (2012) Complete mitochondrial genomes of *Diplogonoporus balaenopterae* and *Diplogonoporus grandis* (Cestoda: Diphyllobothriidae) and clarification of their taxonomic relationships. Parasitol Int 61(2):260–266

Yamasaki H, Tsubokawa D, Mercado R, Kuramochi T (2014) A simple method for identifying the diphyllobothriids based on mitochondrial DNA analysis. In: Takamiya S (ed) Parasitological research: materials and methods. Sankeisha, Tokyo, pp 47–49 (in Japanese)

Yamasaki H, Kumazawa H, Sekikawa Y, Oda R, Hongo I, Tsuchida T et al (2016) First confirmed human case of *Diphyllobothrium stemmacephalum* infection and molecular verification of the synonymy of *Diphyllobothrium yonagoense* with *D. stemmacephalum* (Cestoda: Diphyllobothriidea). Parasitol Int 65(5):412–421

Yamasaki Y, Morishima H, Sugiyama H (2017) Cestode infections in Japan. IASR 38:74–76 (in Japanese)

Yamasaki H, Morishima Y, Sugiyama H, Okamoto M (2021a) Current situation of human *Taenia asiatica* taeniosis in Japan. Parasitol Int (in press)

Yamasaki H, Sanpool O, Rodpai R, Sadaow L, Laummaunwai P, Un M et al (2021b) *Spirometra* species from Asia: Genetic diversity and taxonomic challenges. Parasitol Int 80:102181

Yera H, Nicoulaud J, Dupouy-Camet J (2008) Use of nuclear and mitochondrial DNA PCR and sequencing for molecular identification of *Diphyllobothrium* isolates potentially infective for humans. Parasite 15(3):402–407

Yong TS, Chai JY, Sohn WM, Eom KS, Jeoung HG, Hoang EH et al (2014) Prevalence of intestinal helminths among inhabitants of Cambodia (2006-2011). Korean J Parasitol 52(6):661–666

Yoshikawa M, Ouji Y, Nishifuku M, Ishizaka S, Nawa Y (2010) Sparganosis cases reported in Japan in the recent decade, 2000-2009. Clin Parasitol 21:33–36 (in Japanese)

Yousofi Darani H, Jafari R (2020) Renal echinococcosis; the parasite, host immune response, diagnosis and management. J Infect Dev Ctries 14(5):420–427

Zain SNM, Behnke JM, Lewis JW (2012) Helminth communities from two urban rat populations in Kuala Lumpur, Malaysia. Parasit Vectors 5:47

Zhang X, Cui J, Wei T, Li LY, Jiang J, Lu JC et al (2014) Survey and genetic variation of *Spirometra erinaceieuropaei* sparganum in frogs and snakes from Guangxi of southern China. Trop Biomed 31(4):862–870

Zhang X, Hong X, Liu SN, Jiang P, Zhao SC, Sun CX et al (2020) Large-scale survey of a neglected agent of sparganosis *Spirometra erinaceieuropaei* (Cestoda: Diphyllobothriidae) in wild frogs in China. PLoS Negl Trop Dis 14(2):e0008019

# Chapter 7
# Black Fly Diversity and Impacts on Human Welfare in Southeast Asia

**Pairot Pramual**

**Abstract** Black flies (Diptera: Simuliidae) are insects of medical and veterinary importance because some of them are vectors of human and livestock disease agents. Even without transmission of pathogens, black fly biting can affect human and animal welfare through nuisance and irritation. More than 2300 species of black flies have been recorded globally and about 20% of these occur in Southeast Asia (SE Asia). A total of 456 black fly species arranged in eight subgenera of the genus *Simulium* are found in this region. Although there is a long history (>100 years) of black fly study in some countries in SE Asia, extensive explorations began less than two decades ago. Recent advances in taxonomy using morphological, cytological, and molecular approaches have provided a backbone for other aspects of black fly study in this region. At least seven species are known as human-biters and two are considered pests in northern Thailand. Three black fly species (*Simulium nodosum* Puri, *S. nigrogilvum* Summers and *S. asakoae* Takaoka and Davies) in SE Asia have been found carrying animal-origin filarial parasites including two unidentified species of the genus *Onchocerca* that possibly cause human zoonotic onchocerciasis. At least two species (*S. asakoae* and *S. chumpornense* Takaoka and Kuvangkadilok) are possible vectors of avian parasites of the genus *Leucocytozoon* and *Trypanosoma*. However, further study is required to evaluate the impacts of black flies either directly (i.e. transmission of disease agents to humans) or indirectly (e.g. economic impact). In addition, it is highly likely that areas of high diversity await exploration in Laos, Cambodia, and some parts of Vietnam.

**Keywords** Aquatic insects · Black fly · Diptera · *Onchocerca* · Vector

---

P. Pramual (✉)
Department of Biology, Faculty of Science, Mahasarakham University, Maha Sarakham, Thailand
e-mail: pairot.p@msu.ac.th



T. N. Petney et al. (eds.), *Biodiversity of Southeast Asian Parasites and Vectors causing Human Disease*, Parasitology Research Monographs 14,
https://doi.org/10.1007/978-3-030-71161-0_7

## 7.1 Biology

Black fly (Fig. 7.1) is the common name of the insects belonging to the family Simuliidae of the order Diptera. The family Simuliidae is divided into two subfamilies, Parasimuliinae Smart and Simuliinae Newman. Only one genus belongs to the subfamily Parasimuliinae, namely, *Parasimulium* Malloch, which contains only four species. The remaining extant black fly species are all members of the subfamily Simuliinae, which includes two tribes, Prosimuliini Enderlein and Simuliini Newman. Prosimuliini is comprised of six genera and 144 species. Simuliini is comprised of 23 genera with 2183 extant species (Adler 2020).

Black flies are aquatic insects because their immature stages (i.e. egg, larva, and pupa) inhabit flowing water. They have two adaptations that may explain their requirement for running water: respiratory and food-filtering mechanisms. The black fly larva receives oxygen from water through the permeable cuticle into the trachea system (Crosskey 1990). Flowing water is typically rich in dissolved oxygen, and is, therefore, suitable to the respiratory system of the black fly larva. The larva obtains food with a pair of labral fans that, in flowing water, are used to filter food that is comprised of organic matter, diatoms, and filamentous algae (Crosskey 1990).

Black fly eggs (Fig. 7.2) are ovoid, <0.5 mm in length, white in color initially but turning brown as they age (Adler et al. 2004). The number of eggs produced by females varies greatly from <20 to almost 800. Size and number of eggs are negatively related and so species that lay bigger eggs usually produce a smaller number of them (Malmqvist et al. 2004). The female black fly lays her eggs on various substrates such as rocks, fallen leaves, wood or even plastic bags that are in the flowing water. In tropical regions, black fly eggs usually hatch within 3–6 days but temperate species may take up to 6 months (Crosskey 1990).

The larva of the black fly (Fig. 7.3) is vermiform but expands posteriorly. Body length varies from 5 to 15 mm and color from white or light brown to dark brown or

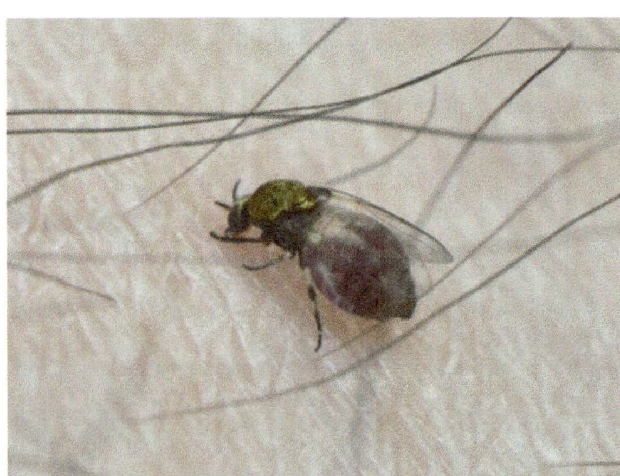

**Fig. 7.1** Female of *Simulium nigrogilvum* Summers biting a human in Chiangmai province, northern Thailand

**Fig. 7.2** Eggs of black fly

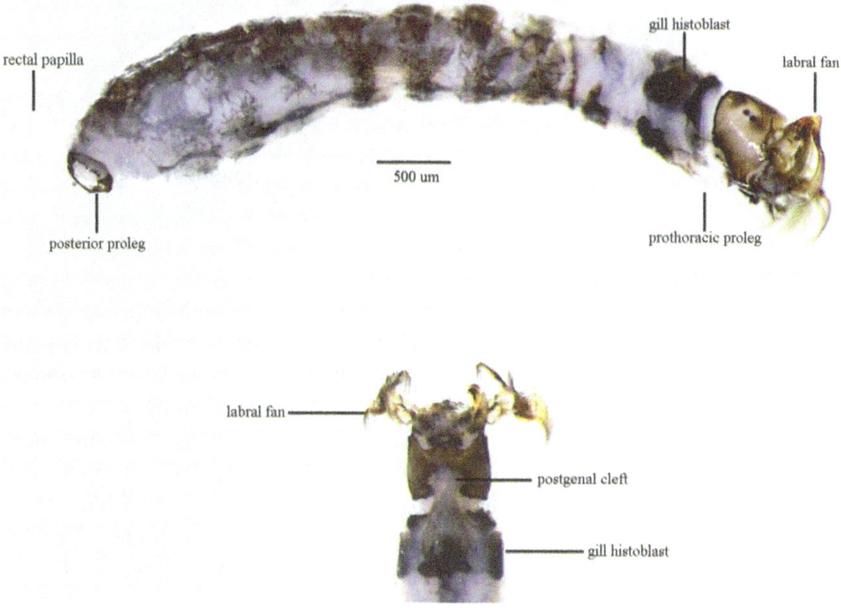


**Fig. 7.3** Larva of black fly, *Simulium nakhonense* Takaoka & Suzuki

dark green. The larval body is divided into three sections: head, thorax, and abdomen. The head is well-sclerotized with a prominent pair of food gathering organs, the labral fans (Fig. 7.3). The ventral area of the head capsule possesses a postgenal cleft (Fig. 7.3), which has uncertain function but is a taxonomically important character because it varies from species to species. The thorax has a

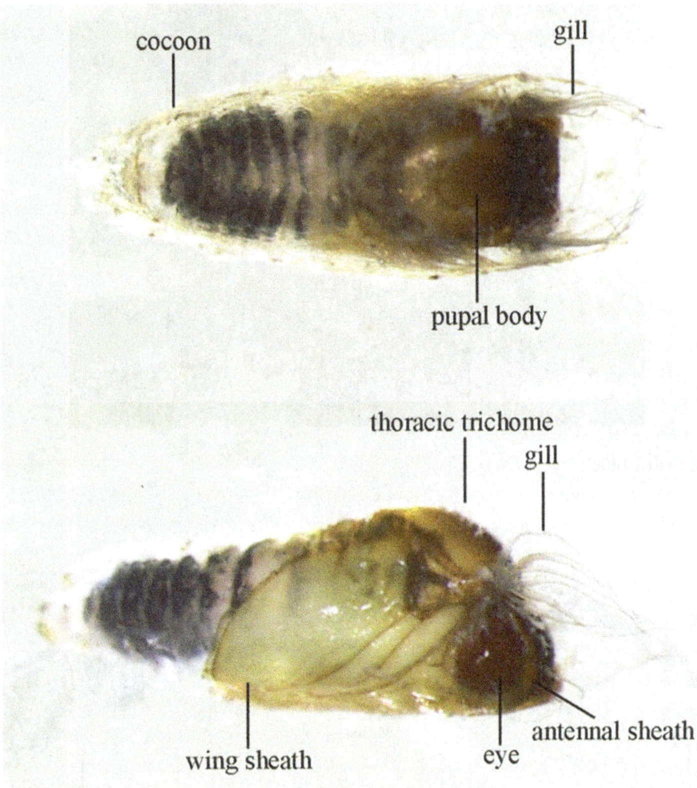


**Fig. 7.4** Pupa of black fly, *Simulium nakhonense*

substrate attachment and locomotion organ, the thoracic proleg (Fig. 7.3) apically bearing rows of tiny hooks used for attaching to silk deposited on a substrate. On the lateral side of the thorax are gill histoblasts (Fig. 7.3) that become the gills (i.e., the respiratory organs) of the pupa. The abdomen is generally wider than other parts because within it there is a large pair of salivary or silk glands. This organ becomes larger with age of the larva as it prepares to make silk for the pupal cocoon. At the posterior end of the abdomen are rows of tiny hooks called the posterior circlet, similar to those of the thoracic proleg, and also used to attach the larva to silk on a substrate. There are from four to nine larval instars although more generally there are seven (Crosskey 1990). The final instar larva is easily recognized by its black gill histoblasts (Fig. 7.3). Development time from first to last instar depends on species and environment. The most important environmental factor affecting larval development is temperature. Species in temperate regions can take months to develop but those in tropical areas generally need only 1–2 weeks to complete the larva stage.

Black fly pupae (Fig. 7.4) have two parts, the pupal body and the cocoon. The pupal body lives within the cocoon and has a shape resembling that of the adult. The pupal body has three parts: head, thorax, and abdomen. There are

**Fig. 7.5** Variation in gills of black flies (cocoons of *S. aureohirtum* and *S. oblongum* have been removed)

prominent structures called "gills" on the lateral sides of the thorax. Pupal gills vary in shape, size, number, and arrangements of filaments (Fig. 7.5) and are, therefore, valuable characters for species identification. The pupal cocoon has two main shapes, slipper-shaped and boot-shaped (Fig. 7.6). There are additional characters of the cocoon that are also taxonomically important including presence/absence of lateral windows or presence/absence of an anterodorsal projection (Fig. 7.6). The pupa develops into an adult within 2–3 days but can take weeks depending on the temperature (Adler et al. 2004).

Immature stages of black flies use a wide range of running water habitats. It is well recognized that different black fly species prefer different stream habitats (McCreadie and Adler 1998). Factors that are associated with species distribution include stream width, depth, velocity, elevation, riparian vegetation, and presence/absence of a water impoundment (Hamada and McCreadie 1999; Hamada et al. 2002; McCreadie et al. 2004; Pramual and Kuvangkadilok 2009; Pramual and Wongpakam 2010). For example, *S. aureohirtum* Brunetti in Thailand is associated with small streams with very low velocity in open areas. In contrast, *S. nakhonense* Takaoka & Suzuki and *S. angulistylum* Takaoka & Davies complex both prefer large and fast-flowing streams with more cover (Pramual and Kuvangkadilok 2009). Many black fly species are habitat specialized and in SE Asia, two main types of habitat specialization are recognized: high elevation and highly calcareous streams. Many species occupy only high-elevation habitats, such as members of subgenus *Montisimulium* Rubtsov and the *S. vernum* species-group of subgenus *Nevermannia*

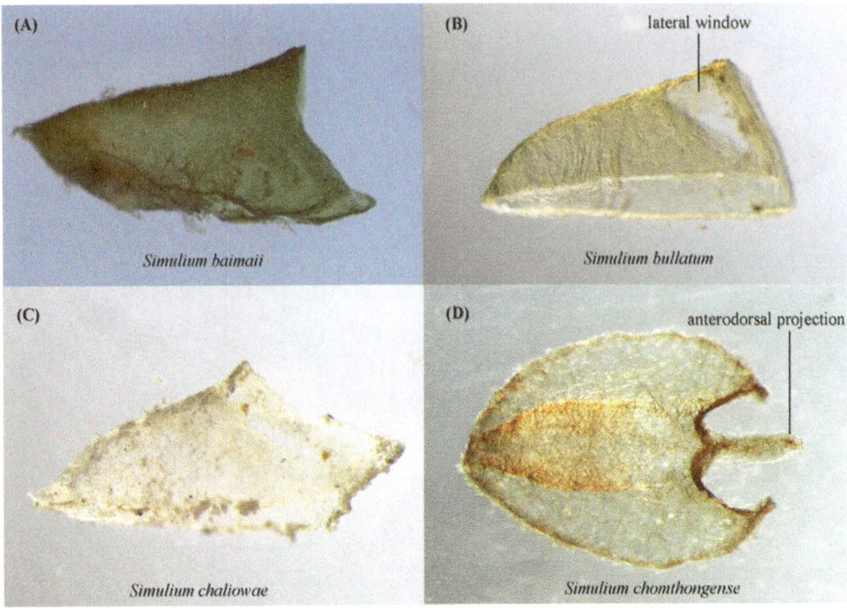


**Fig. 7.6** Cocoon shapes of black fly pupae. (**a**) slipper-shaped in *Simulium baimaii* Kuvangkadilok & Takaoka, (**b**) slipper-shaped with lateral windows in *Simulium bullatum* Takaoka & Choochote, (**c**) boot-shaped in *Simulium chaliowae* Takaoka & Boonkemtong, and (**d**) slipper-shaped with long anterodorsal projection in *Simulium chomthongense* Takaoka, Srisuka & Choochote

Enderlein. Species from Thailand that are restricted to highly calcareous streams include *S. weji* Takaoka, *S. chaliowae* Takaoka and Boonkemtong, *S. lampangense* Takaoka and Choochote, *S. takense* Takaoka and Choochote, and *S. triglobus* Kuvangkadilok and Takaoka (Thaijarern et al. 2018). Population genetic studies have found that black fly species that require specialized habitats show high levels of population genetic structure due to patchy distribution of the suitable habitat (Pramual and Wongpakam 2013; Pramual and Pangjanda 2015).

The adult black fly (Fig. 7.7) has a stout body ranging between 1.2 and 6.0 mm in length (Crosskey 1990; Adler et al. 2004). Most adults have black or dark brown ground color but some species, such as *S. nigrogilvum* Summers (Fig. 7.8) from Thailand and Vietnam, have yellow pollinosity. Males and females can be recognized by differences in their eye facets. Female have eye facets that are evenly sized while males have larger facets on the upper half of the eye and smaller ones on the lower half.

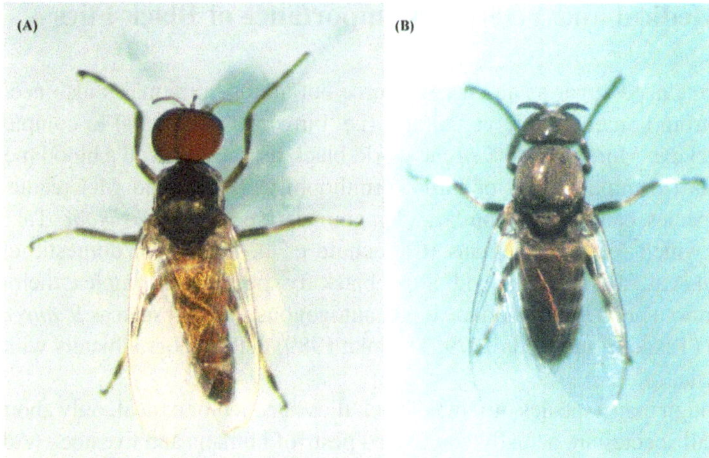


**Fig. 7.7** Adult black flies, *Simulium nakhonense* (**a**) male and (**b**) female (Photograph by Dr. Chaliow Kuvangkadilok)

**Fig. 7.8** *Simulium nigrogilvum* (**a**) larva, (**b**) pupa (cocoon removed), and (**c**) adult female

## 7.2 Medical and Veterinary Importance of Black Flies

Both sexes need sugar as an energy source but almost all females also need blood protein from homeothermic vertebrates (i.e. birds and mammals) to complete their ovarian cycle. More than 90% of the world black fly species need a blood meal from mammals (mammalophily) or birds (ornithophily) (Adler and McCreadie 2019). Some species can feed on birds and mammals, such as *S. asakoae* Takaoka & Davies, which feeds on humans (Choochote et al. 2005) and domestic chickens (Pramual et al. 2020) in Thailand. Some black fly species can complete their ovarian cycle without need for a blood meal (i.e. autogenous species) such as *S. aureohirtum* Brunetti (Takaoka and Noda 1979; Takaoka 1989) that is geographically widespread in SE Asia.

Although many species of female black flies need a blood meal, only about 10%–20% of all species are actually considered pests of humans and livestock (Adler and McCreadie 2019). This biting habit makes them rank third in importance amongst arthropods that transmit disease agents to human and animals (Adler and McCreadie 2019). The most significant disease agent transmitted by black flies is *Onchocerca volvulus*. This filarial nematode is the causative agent of human onchocerciasis (river blindness). At least 27 black fly species or members of species complexes transmit this disease agent (Adler and McCreadie 2019).

Human onchocerciasis was first reported in 1875 on the western coast of Africa and now occurs in 31 countries in Africa and two in South America, although previously it occurred in four South American and two Central American countries (WHO 2019). It affects 20.9 million people who have been infected with *O. volvulus*. Among these, 14.6 million have skin disease and 1.15 million have visual loss (WHO 2019). Recently, this disease was eliminated from Colombia, Mexico, Ecuador, and Guatemala (Rodríguez-Pérez et al. 2015; Guevara et al. 2018; WHO 2019). In addition to human onchocerciasis, black flies can also transmit other *Onchocerca* species among wild and domesticated animals. At least 11 *Onchocerca* species are transmitted to animals such as deer, moose, wild boar, cattle, and reindeer by at least 20 black fly species (Adler and McCreadie 2019). Some of these *Onchocerca* species can occasionally be transmitted to humans and cause a disease termed "zoonotic onchocerciasis" which has been reported in many countries (Takaoka et al. 2012).

In addition to filarial species of the genus *Onchocerca*, black flies are also vectors of three other filarial nematodes, namely *Dirofilaria ursi*, *Mansonella ozzardi*, and *Splendidofilaria fallisensis* (Adler and McCreadie 2019). The *Simulium venustum* Say complex is a vector of *D. ursi* transmitted among bears in North America. At least five species and species complexes of black flies in South America are vectors of *M. ozzardi* transmitted to humans (Adler and McCreadie 2019). This filarial nematode causes a disease called "mansonellosis," which has low pathogenicity and is almost asymptomatic. However, infection of *M. ozzardi* occasionally causes itching, enlarged lymph glands, and joint pains (Ta-Tang et al. 2018). Two black fly species, *S. anatinum* and *S. rugglesi* in North America are vectors of *S. fallisensis* in ducks (Anderson 1968).

In addition to filarial nematodes, black flies also transmit other disease agents such as viruses and protozoa to animals including economically important livestock. Two black fly species, *S. notatum* and *S. vittatum*, in America are vectors of the vesicular stomatitis virus. Cattle, horses, pigs, goats, llama, and sheep are hosts of the vesicular stomatitis virus (Adler and McCreadie 2019). Black flies are major vectors of the avian blood protozoan genus *Leucocytozoon*. Almost 50 species of the genus *Leucocytozoon* have been described but molecular genetic investigations show that diversity is much greater (Fecchio et al. 2020). All of the *Leucocytozoon* species are transmitted by black flies except for *Leucocytozoon* (*Akiba*) *caulleryi* that has a biting midge of the genus *Culicoides* as the vector (Valkiūnas 2005). At least 21 species of black flies have been reported as vectors of *Leucocytozoon* (Adler and McCreadie 2019). Infection of *Leucocytozoon* can cause a disease known as leucocytozoonosis, occurring in both wild and domestic birds. However, effects of the disease vary between protozoa and host species. Co-infection with other blood protozoa of the genera *Plasmodium* and *Haemoproteus*, which are typically transmitted by mosquitoes, can increase virulence of the disease in the hosts (Pigeault et al. 2018).

## 7.3 Biodiversity of Black Flies in Southeast Asia

Southeast Asia (SE Asia) is among the most biodiverse regions in the world. Despite covering only 4% of the Earth's land area, about 20–25% of the global flora and fauna inhabit Southeast Asia (Woodruff 2010). More than 15,000 plant species are endemic to this region. Southeast Asia also ranks second among species richness in freshwater ecosystems (Woodruff 2010). The wealth of biodiversity of SE Asia is a result of complex geological historical, and climatic conditions ranging from subtropical to tropical, and great variation in elevation (0–4101 m of Mt. Kinabalu). This complexity of geology and climatology created a wide variety of ecoregions that are suitable for a diversity of living organisms (Woodruff 2010). Although rich in diversity, this region is one of the most threatened biodiversity hotspots (Myers et al. 2000) because it is also experiencing a high rate of diversity loss (Sodhi et al. 2010).

The family Simuliidae in Southeast Asia has a long history of study. For example, the first black fly species, *S. nigrogilvum*, was reported in Thailand from more than a century ago (Summers 1911). However, extensive taxonomic study began <40 years ago in some countries such as the Philippines (Takaoka 1983) and Thailand (Takaoka and Suzuki 1984; Takaoka and Choochote 2004) and more recently in Malaysia (Takaoka and Davies 1995) and Indonesia (Takaoka and Davies 1996). Study of biodiversity-rich countries, such as Vietnam, started only <5 years ago (Takaoka et al. 2017a) and some countries, such as Cambodia and Laos, are totally unexplored. Despite this limitation, Southeast Asia holds nearly 20% (456 of 2331) of the world black fly species. All black fly species reported in SE Asia thus far belong to eight subgenera of the genus *Simulium* Latreille s. str. (Table 7.1). No other genera have been recorded in the region. Species of two

**Table 7.1** Numbers of black fly species of the genus *Simulium* Latreille s. str. in Southeast Asian countries

| Subgenus | Indonesia | Malaysia | Myanmar | Philippines | Thailand | Vietnam | Total |
|---|---|---|---|---|---|---|---|
| *Asiosimulium* Takaoka & Choochote | 0 | 0 | 1 | 0 | 5 | 0 | 6 |
| *Daviesellum* Takaoka & Adler | 0 | 1 | 0 | 0 | 2 | 0 | 2 |
| *Gomphostilbia* Enderlein | 58 | 50 | 10 | 33 | 37 | 26 | 176 |
| *Montisimulium* Rubtsov | 0 | 0 | 1 | 0 | 6 | 1 | 8 |
| *Morops* Enderlein | 28 | 0 | 0 | 0 | 0 | 0 | 28 |
| *Nevermannia* Enderlein | 9 | 7 | 2 | 4 | 10 | 9 | 34 |
| *Simulium* Latreille s. str. | 44 | 37 | 14 | 37 | 54 | 41 | 186 |
| *Wallacellum* Takaoka | 2 | 0 | 0 | 14 | 0 | 0 | 16 |
| Total | 141 | 95 | 27 | 88 | 114 | 77 | 456 |

Data from Adler et al. (2020)

subgenera, *Simulium* Latreille s. str. and *Gomphostilbia* Enderlein, are the major components of the black fly faunas of SE Asia. These two subgenera make up of nearly 80% (362 from 456 species) of black fly species in this region.

Among eight subgenera found in SE Asia, *Daviesellum* Takaoka & Adler and *Wallacellum* Takaoka are endemic to the region. The former two subgenera are small, with only two and 17 species, respectively. Two additional subgenera (i.e. *Asiosimulium* and *Gomphostilbia*) have the majority of their species endemic to SE Asia. *Asiosimulium* has seven species recorded globally and six of these are found in SE Asia and one is found in Nepal (Adler 2020). Most *Gomphostilbia* (176 from 271 species) occur in SE Asia (Adler 2020).

Black flies are thought to have originated in cool, mountainous environments as members of old lineages such as the genera *Parasimulium* Malloch, *Gymnopais* Stone, *Helodon* Enderlein, and *Prosimulium* Roubaud, which are all restricted to cool habitats in the Northern Hemisphere (Adler et al. 2010). Because no members of these ancestral lineages occur in SE Asia, it has been proposed that black flies in SE Asia were derived from ancestral lineages in the northern areas of the Palearctic region that moved southward during glacial periods (Takaoka 2017). The subgenus *Gomphostilbia* probably evolved and radiated in SE Asia. Two cosmopolitan subgenera, *Simulium* and *Nevermannia* Enderlein, dispersed from the Palearctic during glaciations and extended their distributions southward to Australia. The subgenus *Montisimulium* Rubtsov has a center of distribution in the Palearctic. Eight species are found in Southeast Asia, mainly at high elevations (>1000 m above sea level) and they presumably evolved from an ancestor derived from high-latitude areas in the Palearctic region and then moved southward during the glaciations (Takaoka 2017).

Ten SE Asian black flies were originally discovered to be species complexes (Table 7.2). Another species complex that occurs in SE Asia is the *S. ornatipes* Skuse complex. However, populations of this species complex were cytologically examined only in Australia and associated islands (Bedo 1977, 1979). There is no report on the cytogenetic variation of *S. ornatipes* in Indonesia.

Most studies of cytological species complexes in SE Asia have taken place in Thailand, but there are also some reports from Malaysia and Vietnam. Despite this geographic bias of cytological examinations, high diversity at the chromosomal level has already been highlighted for the black flies in Vietnam (Adler et al. 2016). There are additional examples of hidden diversity. For example, the *S. tani* complex is geographically widespread, having been recorded in Malaysia, Myanmar, Indonesia, Thailand, Taiwan, and Vietnam. Cytological examinations revealed that this nominal species consists of 12 cytoforms (Tangkawanit et al. 2009; Adler et al. 2013, 2016), the largest species complex in the Oriental region. Molecular genetic study has shown that some cytoforms are genetically distinct and probably represent valid species (Low et al. 2016).

Habitat-specialized black fly species, particularly those occupying high elevations, also show high diversity. Populations that use high-elevation habitats are geographically isolated by lowland areas and are thus isolated on "ecological islands," analogous with those occurring on oceanic islands. For example, high

**Table 7.2** Black fly species complexes in Southeast Asia

| Subgenus/ Species | Cytoform | Reference |
|---|---|---|
| *Gomphostilbia* | | |
| *Simulium angulistylum* complex | Three cytoforms (A, B, C), all in Thailand | Pramual and Kuvangkadilok (2012) |
| *Simulium sheilae* complex | Two cytological lineages, all in Thailand | Jitklang et al. (2008) |
| *Simulium siamense* complex | Seven cytoforms (A, B, C, D, E, F, G), all in Thailand | Kuvangkadilok et al. (2008), Pramual and Wongpakam (2011) |
| *Nevermannia* | | |
| *Simulium chomthongense* complex | Two cytoforms (A, B), all in Thailand | Adler et al. (2020) |
| *Simulium feuerborni* complex | Four cytoforms (A, B, C, D): Two cytoforms (A, B) in Thailand, one cytoform (C) in Malaysia and one cytoform (D) in Indonesia | Pramual and Wongpakam (2013), Pramual et al. (2015) |
| *Simulium* | | |
| *Simulium doipuiense* complex | Six cytoforms (A, B, C, D, E, F): two cytoforms (A, B) in Thailand, four cytoforms (C, D, E, F) in Vietnam | Tangkawanit et al. (2009); Adler et al. (2016) |
| *Simulium fenestratum* complex | At least two cryptic species in Thailand based on chromosome polymorphisms. | Thaijarern et al. (2018) |
| *Simulium malayense* complex | Three cytoforms (A, B, C): cytoforms A and B in Thailand and cytoform C in Malaysia. | Thaijarern et al. (2018) |
| *Simulium tani* complex | Twelve cytoforms (A, B, C, D, E, F, G, H, I, K, L, B2): Nine cytoforms (A–I) in Thailand, one cytoform (K) in Malaysia, one (L) cytoform in Taiwan and one (B2) cytoform in Vietnam. | Tangkawanit et al. (2009); Adler et al. (2013, 2016) |
| *Simulium tenebrosum* complex | Two cytoforms (A, B) in Thailand | Adler et al. (2019) |

levels of genetic differentiation at both molecular and chromosome levels have been reported in the *S. feuerborni* complex. This species occurs only in high-elevation habitats at >700 m above sea level (Pramual and Wongpakam 2013; Pramual et al. 2015). It is not only show high genetic divergence between geographically isolated populations, but cryptic diversity at individual high-elevation sites for the *S. tenebrosum* complex (Adler et al. 2019) and *S. chomthongense* complex (Adler et al. 2020), which inhabit the highest mountain (i.e. Doi Inthanon, 2565 m above sea level) in Thailand. Other types of habitat-specialized species, such as those using highly calcareous streams, also show similar patterns with those of high-elevation habitats. Population genetic study in *S. weji*, an endemic species of Thailand that has immature stages associated with highly calcareous streams, revealed a high level of genetic differentiation even between geographically close populations (Pramual and Pangjanda 2015). The results of these studies indicate that there is possibly hidden diversity in many morphologically described species. Given the morphological homogeneity of the Simuliidae, it is, therefore, necessary to use an integrated approach to uncover hidden diversity (e.g. Ilmonen et al. 2009; Adler and Huang 2011; Pramual and Kuvangkadilok 2012; Diaz et al. 2015).

It has been suggested that adaptation to different ecological niches is mediated by chromosomal inversions driving sympatric speciation in black flies (Rothfels 1989). Although it is difficult to know whether speciation is sympatric or allopatric, there is evidence indicating that different cytoforms are ecologically different (Adler and Kim 1984; Adler 1988; Adler and McCreadie 1997; Boakye et al. 1998; Pramual and Kuvangkadilok 2012). For example, the immature stages of two cytoforms (B and C) of *S. angulistylum* complex in Thailand are associated with different habitats. Cytoform B occurs in low (<600 m) elevation habitats, whereas those of the cytoform C are found on high mountains (>1000 m above sea level). These cytoforms are also molecularly different with >3.1% sequence divergence, based on Kimura 2-parameter model (Pramual and Kuvangkadilok 2012). Thus, they are probably different biological species. This indicated that chromosomal variations potentially promote species differentiation through ecological shifts (Kirkpatrick and Barton 2006).

Molecular population genetic and phylogeographic studies of black flies in SE Asia have revealed that historical climatic and environmental changes during the Pleistocene potentially influenced the present day genetic structure and diversity of these insects. During the Pleistocene, there were cycles of cool/dry and warm/humid conditions with approximately 100,000-year intervals. During the cool/dry periods, most stream habitats of immature black flies potentially dried up, resulting in population restriction. Once the climate changed to warm/humid condition, streams began to flow and, thus, populations of black flies began expanding. Phylogeographic studies in some black fly species in Thailand found evidence of population expansion dating back to the Pleistocene (Pramual et al. 2005; Thaijarern et al. 2014). Pleistocene climatic and environmental change could also have been driving species diversification of black flies in Thailand, particularly for species occupying high-elevation habitats (Adler et al. 2019).

## 7.4 Public Health Important of Black Flies in Southeast Asia

There are no reports of human onchocerciasis or other human diseases related to black flies in SE Asia thus far. The main problems associated with black flies in this region are due to the effects of biting. Skin dermatitis as a result of black fly biting is a reaction of the immunoglobulin E (IgE) against salivary gland proteins and other

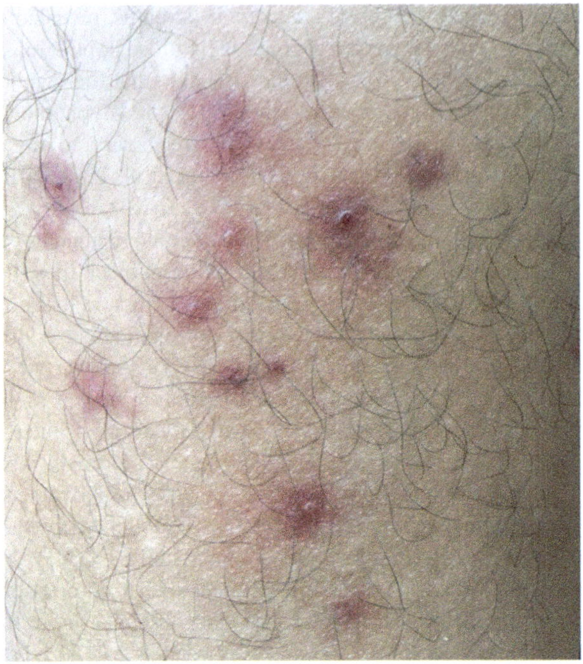

**Fig. 7.9** Reaction on human skin to bites of *Simulium nodosum* in Mae Hong Son province, northern Thailand

active molecules released during biting (Hellberg et al. 2009; Schaffartzik et al. 2009; Hempolchom et al. 2019). Characterization of the IgE-binding proteins in the salivary glands of *S. nigrogilvum*, a black fly pest species in northern Thailand, detected nine proteins that are involved in allergic reactions (Hempolchom et al. 2019). Different people respond differently to the effects of black fly bites. The most common reaction when bitten by black fly is a small red spot with oozing blood followed by swelling (Fig. 7.9), although some people may be allergic and have more serious symptoms, such as vomiting, headache, or fever. Recovery usually follows within a few weeks.

Among the 456 black fly species reported in SE Asia, the human-biting habit is only reported from Thailand. Seven taxa are known to bite humans (Table 7.3), namely, *S. asakoae*, *S. chamlongi* Takaoka & Suzuki, *S. doipuiense* Takaoka & Choochote complex, *S. nigrogilvum*, *S. nodosum*, *S. tenebrosum* Takaoka, Srisuka & Saeung complex, and *S. umphangense* Takaoka, Srisuka & Saeung (Choochote et al. 2005; Pramual et al. 2016; Takaoka et al. 2017b). Two additional species, *S. monglaense* and *S. myanmarense*, originally described from Myanmar were recorded as human-biting species in Thailand based on COI sequence similarity of the specimens morphologically identified as *S. asakoae* (Jomkumsing et al. 2019). *Simulium asakoae* could be readily differentiated from other closely related species

**Table 7.3** Species of black flies in Southeast Asia regarded as potential pests and vectors of disease agents to humans and other animals

| Species | Disease agent | Reference |
|---|---|---|
| Human-biting | | |
| *Simulium asakoae* | Unidentified filarial parasites | Fukuda et al. (2003), Ishii et al. (2008) |
| *Simulium nigrogilvum* | Unidentified filarial parasites *Onchocerca* sp. | Fukuda et al. (2003), Saeung et al. (2020) |
| *Simulium nodosum* | *Onchocerca* sp. | Takaoka et al. (2003), Ishii et al. (2008) |
| *Simulium chamlongi* | – | Choochote et al. (2005), Jomkumsing et al. (2019) |
| *Simulium doipuiense* complex | – | Pramual et al. (2016) |
| *Simulium tenebrosum* complex (reported as *S. rufibasis*) | – | Choochote et al. (2005), Pramual et al. (2016), Takaoka et al. (2019) |
| *Simulium umphangense* | – | Takaoka et al. (2017b) |
| Water buffalo-biting | | |
| *Simulium nodosum* | *Onchocerca* sp. | Takaoka et al. (2003) |
| *Simulium nakhonense* | – | Takaoka et al. (2003) |
| Chicken-biting | | |
| *Simulium asakoae* | *Leucocytozoon* sp. *Trypanosoma* sp. | Jumpato et al. (2019), Thaijarern et al. (2019), Pramual et al. (2020) |
| *Simulium chumpornense* | *Leucocytozoon* sp. *Trypanosoma* sp. | Jumpato et al. (2019), Thaijarern et al. (2019), Pramual et al. (2020) |

Note that all reports are from Thailand

by the larva, pupa, and adult male (Takaoka and Davies 1995; Takaoka et al. 2018a). However, morphological differentiation of females is difficult because they lack reliable diagnostic characters. Recent study has shown that COI barcoding sequences can successfully differentiate closely related species of the *S. asakoae* species-group, including *S. myanmarense* and *S. monglaense* (Low et al. 2020). Thus, COI sequences of some human-biting specimens from northern Thailand originally identified as *S. asakoae* complex were similar to those of *S. myanmarense* and *S. monglaense*, suggesting that these species could also occur in Thailand and are human-biters (Jomkumsing et al. 2019).

Among black fly species that bite humans, only two species, *S. nigrogilvum* (Fig. 7.8) and *S. nodosum* (Fig. 7.10), are considered pests and only in some areas in northern Thailand. *Simulium nigrogilvum* has been recorded in Thailand, Myanmar, and Vietnam (Adler 2020) and was first recognized as a human-biting species in northern Thailand (Takaoka et al. 2003) where it was considered a pest in some areas (Adler and McCreadie 2019). *Simulium nigrogilvum* occurs at mid–high elevations (849–1589 m above sea level) and is a common human-biter at these areas (Choochote et al. 2005; Pramual et al. 2016). Seasonal observations revealed two

**Fig. 7.10** *Simulium nodosum* (**a**) larva, (**b**) pupa, (**c**) male, and (**d**) female. (Male and female photograph by Dr. Chaliow Kuvangkadilok)

peaks of abundance for adult females attracted to humans, one in the dry season (March) and another at the beginning (June) of the rainy season (Choochote et al. 2005). The pattern of adult abundance was thought to be related to the availability of the stream habitat of the immature stages. However, thus far, little is known about the ecology of the immature stages of *S. nigrogilvum*.

*Simulium nodosum* was first reported biting humans in India (Lewis 1974) and later in northern Thailand (Takaoka et al. 2003). In addition to humans, *S. nodosum* also bites buffalo (Takaoka et al. 2003). *Simulium nodosum* has a high rate of biting activity in the cool–dry season (Jan–May), with a peak in March (Takaoka et al. 2003; Ishii et al. 2008). The daily biting rate of this species is mostly unimodal with a main peak in late afternoon (16.00–18.00) (Takaoka et al. 2003). Comparison among human-biting species in Thailand indicates that *S. nodosum* is most abundant in low-elevation areas (Choochote et al. 2005; Pramual et al. 2016). This is consistent with ecological observations of the larval habitat, which found that this species has a limited elevation range, occurring only at around 800 m above sea level

(Tangkawanit et al. 2011). Although *S. nodosum* is among the most abundant human-biting species in lowland areas, geographic distribution of its biting habit is localized. All reports thus far for the human-biting habit of *S. nodosum* are from the northern regions, although large numbers of immature stages were present in western Thailand (Kanchanaburi province) where there are no reports of human-biting (Chaiyasan and Pramual 2016). In addition to Thailand, *S. nodosum* has been reported from Vietnam and Myanmar (Adler 2020). However, no biting habits of this species have been reported in these countries. Therefore, further examinations would be useful for a better understanding of the role of *S. nodosum* as a pest and vector.

Many areas in northern Thailand include popular tourist places such as Doi Inthanon National Park, Chiangmai province, Phu Soi Dao National Park, Uttaradit province, Mae Wong National Park, Kamphaeng Phet province, where black fly biting is common. There has been at least one case of local people in Huai Mo village, Doi Saket district, Chiangmai province, needing to visit the hospital because of a serious allergic reaction to black fly biting. Economically, black fly biting can cause direct and indirect costs. People living in areas where biting is common need some kind of personal protection and anti-itching medicines (Adler et al. 2004). Black fly biting can reduce the number of tourists in the national parks, thus decreasing income to the parks and also to local people who depend on the tourism industry. Estimates in other regions indicate that black fly outbreaks can cause losses of several million US dollars (Adler et al. 2004; Sarıözkan et al. 2014). However, there is no formal study of the economic impact of black fly biting in Thailand or other countries in SE Asia.

In addition to the role as pests of humans and other animals such as cattle, buffalo, and domestic chicken, black flies in SE Asia are potentially also vectors of human disease agents (Table 7.3). Three black fly species, *S. asakoae*, *S. nigrogilvum*, and *S. nodosum*, were reported as vectors of filarial parasites in Thailand (Takaoka et al. 2003; Fukuda et al. 2003; Ishii et al. 2008). The filarial larvae of a possible *Onchocerca* sp. was detected in wild adult females of *S. nodosum* and *S. nigrogilvum*. The filarial larvae found in *S. nodosum* in northern Thailand are similar to those reported infecting cattle in Japan, suggesting the possibility that this black fly species is a vector of the *Onchocerca* sp. transmitted among cattle and buffalos in Thailand (Takaoka et al. 2003; Ishii et al. 2008). Two filarial species were found in *S. nigrogilvum* in Thailand. One adult specimen from Chiang Mai province, northern Thailand, had an unknown filarial species of *Onchocerca* or a genus of Dirofilariinae (Fukuda et al. 2003). Four specimens of *S. nigrogilvum* from Tak province, western Thailand, had larvae of *Onchocerca* sp. (Saeung et al. 2020). In addition, a filarial larva, possibly of bird origin, in the subfamily Splendidofilariinae or Lemdaninae was also detected in *S. asakoae* (Ishii et al. 2008).

Detection of filarial parasites, possibly of the genus *Onchocerca*, in at least two human-biting black fly species in Thailand suggested that the flies potentially transmit these parasites to humans. The unidentified *Onchocerca* spp. found in *S. nodosum* and *S. nigrogilvum* in Thailand are most likely transmitted among animals. Occasionally these parasites are transmitted to humans and can develop into zoonotic onchocerciasis (Takaoka et al. 2012). There are 37 cases of zoonotic onchocerciasis reported globally (Saeung et al. 2020). Although zoonotic onchocerciasis has not been reported in SE Asia, the vector species have been postulated (i.e. *S. nigrogilvum*, *S. nodosum*) (Takaoka et al. 2003; Fukuda et al. 2003; Ishii et al. 2008; Saeung et al. 2020), thus there is high possibility of the disease can be found in humans.

Although black fly research in SE Asia started more than a century ago, the last two decades have seen more extensive studies. Species checklists along with morphological keys (Hadi and Takaoka 2018; Takaoka et al. 2017a, 2018b, 2019) and DNA barcode databases (e.g. Pramual et al. 2011, 2016; Pramual and Adler 2014; Jomkumsing et al. 2019; Low et al. 2020) have expanded rapidly in this region. This recent strong taxonomic works make SE Asia among the most advanced in black fly taxonomy. Taxonomic information provides an important backbone for other research. Detection of filarial parasites (Takaoka et al. 2003; Fukuda et al. 2003; Ishii et al. 2008; Saeung et al. 2020) and avian blood protozoa (Table 7.3) (Jumpato et al. 2019; Thaijarern et al. 2019; Pramual et al. 2020) highlight the possible significance of these insects as vectors of disease agents in both humans and livestock. However, information regarding pest and vector roles is limited to Thailand. There is a need for further investigation of these roles in countries (e.g. Malaysia, Myanmar, and Vietnam) where pest and vector species such as *S. nodosum*, *S. nigrogilvum*, and *S. asakoae* occur, and will be useful for understanding the regional impacts on human welfare. Potentially species-rich countries such as Laos and Cambodia, which thus far are totally unexplored, will increase knowledge of species diversity of black flies in SE Asia.

**Acknowledgements** I would like to thank Adrian Plant, Peter Adler, Hiroyuki Takaoka, and Ubon Tangkawanit for comments on earlier version of the manuscript.

## References

Adler PH (1988) Ecology of black fly sibling species. In: Kim KC, Merritt RW (eds) Black flies: ecology, population management, and annotated world list. Pennsylvania State University Press, University Park, pp 63–76

Adler PH (2020) World blackflies (Diptera: Simuliidae): a comprehensive revision of the taxonomic and geographical inventory [2020]. Available via DIALOG. https://biomia.sites.clemson.edu/pdfs/blackflyinventory.pdf. Accessed 12 Apr 2020

Adler PH, Huang YT (2011) Integrated systematics of the Simuliidae (Diptera): evolutionary relationships of the little-known Palearctic black fly *Simulium acrotrichum*. Can Entomol 143:612–628

Adler PH, Kim KC (1984) Ecological characterization of two sibling species, IIIL-1 and IS-7, in the *Simulium vittatum* complex (Diptera: Simuliidae). Can J Zool 62:1308–1315

Adler PH, McCreadie JW (1997) The hidden ecology of black flies: sibling species and ecological scale. Am Entomol 43:153–161

Adler PH, McCreadie JW (2019) Black flies (Simuliidae). In: Mullen GR, Durden LA (eds) Medical and veterinary entomology, 3rd edn. Elsevier, San Diego, CA, pp 237–259

Adler PH, Currie DC, Wood DM (2004) The black flies (Simuliidae) of North America. Cornell University Press, Ithaca, NY

Adler PH, Cheke RA, Post RJ (2010) Evolution, epidemiology, and population genetics of black flies (Diptera: Simuliidae). Infect Genet Evol 10(7):846–865

Adler PH, Huang YT, Reeves WK, Kim SK, Otsuka Y, Takaoka H (2013) Macrogenomic evidence for the origin of the black fly *Simulium suzukii* (Diptera: Simuliidae) on Okinawa Island, Japan. PLoS One 8(8):e70765

Adler PH, Takaoka H, Sofian-Azirun M, Low VL, Ya'cob Z, Chen CD, Lau KW, Da Pham X (2016) Vietnam, a hotspot for chromosomal diversity and cryptic species in black flies (Diptera: Simuliidae). PLoS One 11(10):e0163881

Adler PH, Srisuka W, Low VL, Takaoka H, Saeung A (2019) High-elevation chromosomal diversity of black flies (Diptera: Simuliidae) in Thailand. Insect Syst Diver 3(3):1

Adler PH, Srisuka W, Saeung A (2020) Cryptic species of black flies (Diptera: Simuliidae) at high elevations in the oriental region: the *Simulium vernum* species group in Thailand. Acta Trop 205:105393

Anderson RC (1968) The simuliid vectors of *Splendidofilaria fallisensis* of ducks. Can J Zool 46 (3):610–611

Bedo DG (1977) Cytogenetics and evolution of *Simulium ornatipes* skuse (Diptera: Simuliidae) I. Sibling speciation. Chromosoma 64(1):37–65

Bedo DG (1979) Cytogenetics and evolution of *Simulium ornatipes* Skuse (Diptera: Simuliidae). III. Geographic variation of chromosomal polymorphisms and species divergence. Evolution 33:309–318

Boakye DA, Back C, Fiasorgbor GK, Sib APP, Coulibaly Y (1998) Sibling species distributions of the *Simulium damnosum* complex in the West African Onchocerciasis Control Programme area during the decade 1984–93, following intensive larviciding since 1974. Med Vet Entomol 12:345–358

Chaiyasan P, Pramual P (2016) Population genetic structure and demographic history of the black fly vector, *Simulium nodosum* in Thailand. Med Vet Entomol 30(3):286–292

Choochote W, Takaoka H, Fukuda M, Otsuka Y, Aoki C, Eshima N (2005) Seasonalabundance and daily flying activity of black flies (Diptera: Simuliidae) attracted to human baits in Doi Inthanon national park, northern Thailand. Med Entomol Zool 56:335–348

Crosskey RW (1990) The natural history of black flies. Wiley, London

Diaz SA, Moncada LI, Murcia CH, Lotta IA, Matta NE, Adler PH (2015) Integrated taxonomy of a new species of black fly in the subgenus *Trichodagmia* (Diptera: Simuliidae) from the Páramo Region of Colombia. Zootaxa 3914:541–557

Fecchio A, Chagas CR, Bell JA, Kirchgatter K (2020) Evolutionary ecology, taxonomy, and systematics of avian malaria and related parasites. Acta Trop 204:105364

Fukuda M, Choocote W, Bain O, Aoki C, Takaoka H (2003) Natural infections with filarial larvae in two species of black flies (Diptera: Simuliidae) in northern Thailand. J Trop Med Hyg 31 (2):99–102

Guevara Á, Lovato R, Proaño R, Rodriguez Perez MA, Unnasch T, Cooper PJ, Guderian RH (2018) Elimination of onchocerciasis in Ecuador: findings of post-treatment surveillance. Parasite Vector 11(1):265

Hadi UK, Takaoka H (2018) The biodiversity of black flies (Diptera: Simuliidae) in Indonesia. Acta Trop 185:133–137

Hamada N, McCreadie JW (1999) Environmental factors associated with the distribution of *Simulium perflavum* (Diptera: Simuliidae) among streams in Brazilian Amazonia. Hydrobiologia 397:71–78

Hamada N, McCreadie JW, Adler PH (2002) Species richness and spatial distribution of blackflies (Diptera: Simuliidae) in streams of Central Amazonia, Brazil. Freshw Biol 47(1):31–40

Hellberg W, Mellor PS, Torsteinsdóttir S, Marti E (2009) Insect bite hypersensitivity in the horse: comparison of IgE-binding proteins in salivary gland extracts from *Simulium vittatum* and *Culicoides nubeculosus*. Vet Immunol Immunopathol 132:62–67

Hempolchom C, Sookrung N, Srisuka W, Reamtong O, Sakolvaree Y, Chaicumpa W, Dedkhad W, Jariyapan N, Takaoka H, Saeung A (2019) Characterization of IgE-binding proteins in the salivary glands of *Simulium nigrogilvum* (Diptera: Simuliidae). Parasitol Res 118(8):2353–2359

Ilmonen J, Adler PH, Malmqvist B, Cywinska A (2009) The *Simulium vernum* group (Diptera: Simuliidae) in Europe: multiple character sets for assessing species status. Zool J Linnean Soc 156:847–863

Ishii Y, Choochote W, Bain O, Fukuda M, Otsuka Y, Takaoka H (2008) Seasonal and diurnal biting activities and zoonotic filarial infections of two *Simulium* species (Diptera: Simuliidae) in northern Thailand. Parasite 15(2):121–129

Jitklang S, Kuvangkadilok C, Baimai V, Takaoka H, Adler PH (2008) Cytogenetics and morphotaxonomy of the *Simulium* (*Gomphostilbia*) *ceylonicum* species group (Diptera: Simuliidae) in Thailand. Zootaxa 1917(1):1–28

Jomkumsing P, Tangkawanit U, Wongpakam K, Pramual P (2019) Who is biting you? DNA barcodes reveal cryptic diversity in human-biting black flies (Diptera: Simuliidae). Acta Trop 196:22–29

Jumpato W, Tangkawanit U, Wongpakam K, Pramual P (2019) Molecular detection of *Leucocytozoon* (Apicomplexa: Haemosporida) in black flies (Diptera: Simuliidae) from Thailand. Acta Trop 190:228–234

Kirkpatrick M, Barton N (2006) Chromosome inversions, local adaptation and speciation. Genetics 173(1):419–434

Kuvangkadilok C, Lualon U, Baimai V (2008) Cytotaxonomy of *Simulium siamense* Takaoka and Suzuki (Diptera: Simuliidae) in Thailand. Genome 51(12):972–987

Lewis DJ (1974) Man-biting Simuliidae (Diptera) of northern India. Isr J Entomol 9:23–53

Low VL, Takaoka H, Pramual P, Adler PH, Ya'cob Z, Huang YT, Da Pham X, Ramli R, Chen CD, Wannaket A, Sofian-Azirun M (2016) Delineating taxonomic boundaries in the largest species complex of black flies (Simuliidae) in the oriental region. Sci Rep 6(1):1–8

Low VL, Srisuka W, Saeung A, Tan TK, Ya'cob Z, Yeong YS, Takaoka H (2020) DNA barcoding of *Simulium asakoae* (Diptera: Simuliidae) from Northern Thailand. J Med Entomol 57 (5):1675–1678

Malmqvist B, Adler PH, Strasevicius D (2004) Testing hypotheses on egg number and size in black flies (Diptera: Simuliidae). J Vector Ecol 29(2):248–256

McCreadie JW, Adler PH (1998) Scale, time, space, and predictability: species distributions of preimaginal black flies (Diptera: Simuliidae). Oecologia 114(1):79–92

McCreadie JW, Hamada N, Grillet ME (2004) Spatial-temporal distribution of preimaginal black-flies in Neotropical streams. Hydrobiologia 513:183–196

Myers N, Mittermeier RA, Mittermeier CG, Da Fonseca GA, Kent J (2000) Biodiversity hotspots for conservation priorities. Nature 403(6772):853–858

Pigeault R, Cozzarolo CS, Choquet R, Strehler M, Jenkins T, Delhaye J, Bovet L, Wassef J, Glaizot O, Christe P (2018) Haemosporidian infection and co-infection affect host survival and reproduction in wild populations of great tits. Int J Parasitol Parasites 48(14):1079–1087

Pramual P, Adler PH (2014) DNA barcoding of tropical black flies (Diptera: Simuliidae) of Thailand. Mol Ecol Resour 14(2):262–271

Pramual P, Kuvangkadilok C (2009) Agricultural land use and black fly (Diptera, Simuliidae) species richness and species assemblages in tropical streams, Northeastern Thailand. Hydrobiologia 625(1):173–184

Pramual P, Kuvangkadilok C (2012) Integrated cytogenetic, ecological, and DNA barcode study reveals cryptic diversity in *Simulium* (*Gomphostilbia*) *angulistylum* (Diptera: Simuliidae). Genome 55(6):447–458

Pramual P, Pangjanda S (2015) Effects of habitat specialization on population genetic structure of black fly *Simulium weji* Takaoka (Diptera: Simuliidae). J Asia Pac Entomol 18(1):33–37

Pramual P, Wongpakam K (2010) Seasonal variation of black fly (Diptera: Simuliidae) species diversity and community structure in tropical streams of Thailand. Entomol Sci 13(1):17–28

Pramual P, Wongpakam K (2011) Cytogenetics of *Simulium siamense* Takaoka and Suzuki, 1984 (Diptera: Simuliidae) in northeastern Thailand. Aquat Insects 33(2):171–184
Pramual P, Wongpakam K (2013) Population genetics of the high elevation black fly *Simulium (Nevermannia) feuerborni* Edwards in Thailand. Entomol Sci 16(3):298–308
Pramual P, Kuvangkadilok C, Baimai V, Walton C (2005) Phylogeography of the black fly *Simulium tani* (Diptera: Simuliidae) from Thailand as inferred from mtDNA sequences. Mol Ecol 14(13):3989–4001
Pramual P, Wongpakam K, Adler PH (2011) Cryptic biodiversity and phylogenetic relationships revealed by DNA barcoding of oriental black flies in the subgenus *Gomphostilbia* (Diptera: Simuliidae). Genome 54(1):1–9
Pramual P, Thaijarern J, Sofian-Azirun M, Ya'cob Z, Hadi UK, Takaoka H (2015) Cytogenetic and molecular evidence of additional cryptic diversity in high elevation black fly *Simulium feuerborni* (Diptera: Simuliidae) populations in Southeast Asia. J Med Entomol 52(5):829–836
Pramual P, Thaijarern J, Wongpakam K (2016) DNA barcoding of human-biting black flies (Diptera: Simuliidae) in Thailand. Acta Trop 164:33–40
Pramual P, Thaijarern J, Tangkawanit U, Wongpakam K (2020) Molecular identification of blood meal sources in black flies (Diptera: Simuliidae) suspected as *Leucocytozoon* vectors. Acta Trop 205:105383
Rodríguez-Pérez MA, Fernández-Santos NA, Orozco-Algarra ME, Rodríguez Atanacio JA, Domínguez-Vázquez A, Rodríguez-Morales KB, Real-Najarro O, Prado-Velasco FG, Cupp EW, Richards FO Jr, Hassan HK, González Roldán JF, Kuri-Molares PA, Unnasch TR (2015) Elimination of onchocerciasis from Mexico. PLoS Negl Trop Dis 9(7):e0003922
Rothfels K (1989) Speciation in black flies. Genome 32(4):500–509
Saeung A, Srisuka W, Aupalee K, Fukuda M, Otsuka Y, Taai K, Maleewong W, Takaoka H (2020) Natural infections with larvae of *Onchocerca* species type I in the human-biting black fly, *Simulium nigrogilvum* (Diptera: Simuliidae), in western Thailand. Acta Trop 204:105344
Sarıözkan S, A Inci A, Yıldırım A, Düzlü O, Gray EW, Adler PH (2014) Economic losses during an outbreak of *Simulium* (*Wilhelmia*) species (Diptera: Simuliidae) in the Cappadocia region of Turkey. Turkiye Parazitol Derg 38:116–119
Schaffartzik A, Weichel M, Crameri R, Björnsdóttir S, Prisi C, Rhyner C, Torsteinsdóttir E, Marti E (2009) Cloning of IgE-binding proteins from *Simulium vittatum* and their potential significance as allergens for equine insect bite hypersensitivity. Vet Immunol Immunopathol 132(1):68–77
Sodhi NS, Posa MRC, Lee TM, Bickford D, Koh LP, Brook BW (2010) The state and conservation of southeast Asian biodiversity. Biodivers Conserv 19(2):317–328
Summers SLM (1911) Notes from the Entomological Department of the London School of Tropical Medicine. No. II. Description of a new species of *Simulium* from Siamese hills. Ann Mag Nat Hist 8:586–588
Takaoka H (1983) The blackflies (Diptera: Simuliidae) of the Philippines. Japan Society for the Promotion of Science, Tokyo
Takaoka H (1989) Further observations on the autogeny of *Simulium aureohirtum* Brunetti (Diptera: Simuliidae) in the Ryukyu Islands. Med Entomol Zool 40(1):21–28
Takaoka H (2017) Speciation, faunal affinities and geographical dispersal of black flies (Diptera: Simuliidae) in the oriental region. Acta Trop 166:234–240
Takaoka H, Choochote W (2004) A list of and keys to black flies (Diptera: Simuliidae) in Thailand. Trop Med Health 32:189–197
Takaoka H, Davies DM (1995) The black flies (Diptera: Simuliidae) of West Malaysia. Kyushu University Press, Fukuoka
Takaoka H, Davies DM (1996) The blackflies (Diptera: Simuliidae) of Java Indonesia. Bishop Museum Bulletin in Entomology, Hawaii
Takaoka H, Noda S (1979) Autogeny of the black fly *Simulium* (*Eusimulium*) *aureohirtum* (Diptera: Simuliidae). J Med Entomol 15(2):183–184
Takaoka H, Suzuki H (1984) The black flies (Diptera: Simuliidae) from Thailand. Jap J Sanit Zool 35:7–45

Takaoka H, Choochote W, Aoki C, Fukuda M, Bain O (2003) Black flies (Diptera: Simuliidae) attracted to humans and water buffalos and natural infections with filarial larvae, probably *Onchocerca* sp., in northern Thailand. Parasite 10(1):3–8

Takaoka H, Fukuda M, Otsuka Y, Aoki C, Uni S, Bain O (2012) Blackfly vectors of zoonotic onchocerciasis in Japan. Med Vet Entomol 26(4):372–378

Takaoka H, Sofian-Azirun M, Ya'Cob Z, Chen CD, Lau KW, Low VL, Da Pham X, Adler PH (2017a) The black flies (Diptera: Simuliidae) of Vietnam. Zootaxa 4261(1):1–165

Takaoka H, Srisuka W, Saeung A (2017b) A new human-biting black fly species of *Simulium* (*Simulium*) (Diptera: Simuliidae) from Thailand. J Med Entomol 54(4):945–948

Takaoka H, Srisuka W, Maleewong W, Saeung A (2018a) New records of 13 species of black flies (Diptera: Simuliidae) from Myanmar. Acta Trop 178:281–289

Takaoka H, Ya'Cob Z, Sofian-Azirun M (2018b) Classification, annotated list and keys for the black flies (Diptera: Simuliidae) of Peninsular Malaysia. Zootaxa 4498(1):1–65

Takaoka H, Srisuka W, Saeung A (2019) Checklist and keys for the black flies (Diptera: Simuliidae) of Thailand. Med Entomol Zool 70(2):53–77

Tangkawanit U, Kuvangkadilok C, Baimai V, Adler PH (2009) Cytosystematics of the *Simulium tuberosum* group (Diptera: Simuliidae) in Thailand. Zool J Linnean Soc 155(2):289–315

Tangkawanit U, Kuvangkadilok C, Trinachartvanit W, Baimai V (2011) Cytotaxonomy, morphology and ecology of the *Simulium nobile* species group (Diptera: Simuliidae) in Thailand. Cytogenet Genome Res 134(4):308–318

Ta-Tang TH, Crainey JL, Post RJ, Luz SL, Rubio JM (2018) Mansonellosis: current perspectives. Res Rep Trop Med 9:9

Thaijarern J, Sopaladawan PN, Wongpakam K, Pramual P (2014) Phylogeography of the black fly *Simulium aureohirtum* (Diptera: Simuliidae) in Thailand. Genome 57(8):411–417

Thaijarern J, Adler PH, Pramual P (2018) Limited differentiation among black flies in the *Simulium multistriatum* species group (Diptera: Simuliidae) in Thailand: cryptic species, homosequential species and homosequential cryptic species. Zool J Linnean Soc 184(4):1024–1054

Thaijarern J, Tangkawanit U, Wongpakam K, Pramual P (2019) Molecular detection of *Trypanosoma* (Kinetoplastida: Trypanosomatidae) in black flies (Diptera: Simuliidae) from Thailand. Acta Trop 200:105196

Valkiūnas G (2005) Avian malaria parasites and other haemosporidia. CRC Press, Boca Raton, FL

WHO (2019) Onchocerciasis. Available via DIALOG. https://www.who.int/news-room/fact-sheets/detail/onchocerciasis. Accessed 2 June 2020

Woodruff DS (2010) Biogeography and conservation in Southeast Asia: how 2.7 million years of repeated environmental fluctuations affect today's patterns and the future of the remaining refugial-phase biodiversity. Biodivers Conserv 19(4):919–941

# Chapter 8
# Ticks: A Largely Unexplored Factor in Disease Transmission

**Weerachai Saijuntha, Trevor N. Petney, Ross H. Andrews, and Richard G. Robbins**

**Abstract** Tick species occur on most reptiles, birds, and mammals, including humans. They are major vectors of viral, bacterial, and protozoan pathogens of animals and humans worldwide. There is currently increasing interest in the tick species and the pathogens that ticks transmit in Southeast Asia, although substantial gaps in our knowledge remain. Here we provide an overview of the tick fauna of mainland and insular Southeast Asia, the pathogens that the ticks transmit, and likely future changes in the structure of tick communities.

**Keywords** Ixodidae · Argasidae · Southeast Asia · Biodiversity · Pathogens

---

W. Saijuntha (✉)
Walai Rukhavej Botanical Research Institute, Biodiversity and Conservation Research Unit, Mahasarakham University, Maha Sarakham, Thailand
e-mail: weerachai.s@msu.ac.th

T. N. Petney
Departments of Zoology and Paleontology and Evolution, State Museum of Natural History, Karlsruhe, Germany

Cholangiocarcinoma Research Institute, Khon Kaen University, Khon Kaen, Thailand

R. H. Andrews
Cholangiocarcinoma Research Institute, Faculty of Medicine, Khon Kaen University, Khon Kaen, Thailand

Faculty of Medicine, Imperial College, London, UK

R. G. Robbins
Emeritus Civilian Medical Entomologist at the Walter Reed Biosystematics Unit, Department of Entomology, Smithsonian Institution, MSC, MRC 534, Suitland, MD, USA

 
T. N. Petney et al. (eds.), *Biodiversity of Southeast Asian Parasites and Vectors causing Human Disease*, Parasitology Research Monographs 14,
https://doi.org/10.1007/978-3-030-71161-0_8

## 8.1 Introduction

Formal tick research in Southeast Asia dates from the final years of Western colonialism, which yielded seminal works by Toumanoff (1944, Vietnam), Anastos (1950, Indonesia), and Kohls (1950, the Philippines; 1957, Malaysia), to cite some noteworthy examples. It then underwent a period of minimal interest prior to the Vietnam War, when both American and Russian workers conducted major surveys in a number of Southeast Asian countries (Petney et al. 2007). This was followed by continuing efforts, particularly by Harry Hoogstraal and coworkers, until Hoogstraal's untimely death in 1986 (Petney et al. 2007). Recently there has been an upsurge of interest in both the taxonomy of Southeast Asian ticks (Apanaskevich and coworkers) and the pathogens that these ticks may transmit (Ahantarig et al. 2008; Petney et al. 2019; Low et al. 2020; Sharifah et al. 2020). In this chapter we summarize current information on tick biodiversity in Southeast Asia, the role of ticks as disease vectors, and the rapidly changing dynamics of tick ecology in this part of the world.

Southeast Asia has been identified as a region with a higher extinction potential for many species than most other areas worldwide (Schipper et al. 2008; Hughes 2017); indeed, this region is undergoing a biodiversity crisis (Sodhi et al. 2004, 2010; Bickford et al. 2012). There are various factors responsible for this situation, predominantly habitat destruction and fragmentation (Crooks et al. 2017; Tölle et al. 2017; Imai et al. 2018); for example, the massive increase in oil palm and rubber plantations at the expense of natural forest areas (Ziegler et al. 2009; Yaap et al. 2010; Gatti et al. 2019), subsistence hunting, the exotic pet industry, and the Oriental "pharmaceutical" industry (Gray et al. 2017), all of which are exacerbated by rapid increases in the human population (Bickford et al. 2012).

Many of the wild reptile, bird and mammal hosts of parasites, including ticks, have been reduced in numbers and their populations restricted to fragmented habitats, increasing the risk of local extinction (Bordes et al. 2015; Fig. 8.1). Thus, some species of ticks with a limited host range are vulnerable to extinction along with their

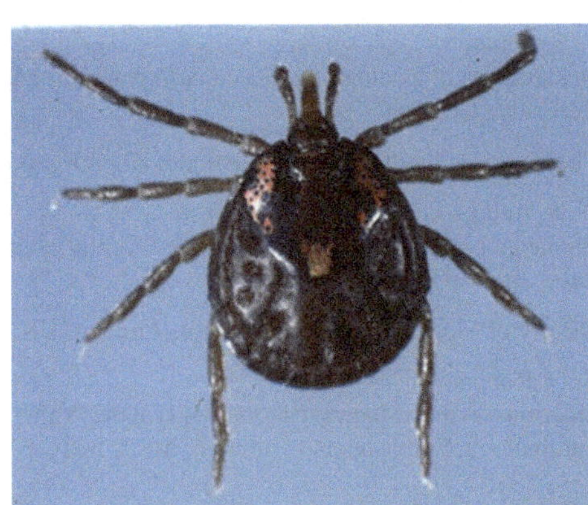

**Fig. 8.1** A female *Amblyomma supinoi* from Myanmar. This is a relatively rare, endangered species found predominantly on tortoises in mainland Southeast Asia. Photo courtesy of James L. Occi, Center for Vector Biology, Rutgers University, New Jersey, USA (see Robbins and Platt 2001)

**Table 8.1** Endangered tick species in Southeast Asia. Modified and enlarged after Durden and Keirans (1996) and Mihalca et al. (2011)

| Ixodidae | |
|---|---|
| *Amblyomma* | *babirussae* |
| | *clypeolatum* |
| | *crenatum* |
| | *javanense* |
| | *komodoense* |
| | *robinsoni* |
| | *supinoi* |
| *Haemaphysalis* | *borneata* |
| | *capricornis* |
| | *kadarsani* |
| | *palawanensis* |
| | *psalistos* |
| | *vietnamensis* |

hosts. Durden and Keirans (1996) list six species (from a total of 48, 12.5%) from Southeast Asia as endangered, while Mihalca et al. (2011) list 10 (from 63, 15.9%) as co-endangered with their hosts (Table 8.1). Both estimates will probably have to be increased as escalating land use changes lead to destruction of natural habitats (Zhao et al. 2006).

By contrast, the increase in stock numbers (particularly cattle, Smith et al. 2018) has substantially increased the number of potential hosts for ticks, and tick-borne diseases, at least for those tick species that are able to utilize these animals as hosts. The large numbers of stray dogs (Traub et al. 2015) also provide suitable hosts for a variety of tick species capable of pathogen transmission (Irwin and Jefferies 2004; Petney et al. 2019).

## 8.2 The Tick Fauna

Southeast Asia has a substantial tick fauna. Petney et al. (2019) listed 97 species from continental Southeast Asia, to which a number of new or reinstated species have been added (*Dermacentor laothaiensis* and *D. pasteuri*, both found in Lao PDR and Thailand, Apanaskevich et al. 2019, 2020) (Table 8.2).

There is no similar summary of species for insular Southeast Asia; however, a list of species known only from islands within the region is provided in Table 8.3. A large number of endemic species occur in this area, some of which are found on specific islands or small island groups; for example, *Amblyomma robinsoni* and *A. komodoense* parasitize Komodo dragons (Guglielmone et al. 2014), large varanid lizards that are found on five small islands within the Lesser Sunda Islands of Indonesia (Ciofi et al. 1999).

In general, the tick fauna of Southeast Asia is strongly biased toward the genus *Haemaphysalis*, followed by substantial numbers of *Amblyomma* and *Dermacentor*.

**Table 8.2** Number of species per tick genus in continental Southeast Asia (modified after Petney et al. 2019)

| Family/genus | Number of species |
|---|---|
| Argasidae | |
| *Argas* | 3 |
| *Ornithodoros* | 2 |
| Ixodidae | |
| *Amblyomma* | 17 |
| *Dermacentor* | 12 |
| *Haemaphysalis* | 43 |
| *Hyalomma* | 2 |
| *Ixodes* | 14 |
| *Nosomma* | 1 |
| *Rhipicephalus* | 6 |
| Total | 101 |

**Table 8.3** Tick species found only on islands in Southeast Asia

| Family/genus | Species |
|---|---|
| Ixodidae | |
| *Amblyomma* (4) | *komodoense* |
| | *kraneveldi* |
| | *robinsoni* |
| | *soembawense* |
| *Dermacentor* (1) | *confragus* |
| *Haemaphysalis* (13) | *bartelsi* |
| | *borneata* |
| | *celebensis* |
| | *hirsuta* |
| | *kadarsani* |
| | *luzonensis* |
| | *mjoebergi* |
| | *psalistos* |
| | *renschi* |
| | *rusae* |
| | *sumatraensis* |
| | *susphilippensis* |
| | *toxopei* |
| *Ixodes* (2) | *collocaliae* |
| | *cordifer* |
| Argasidae | |
| *Ornithodoros* | *collocaliae* |

The genus *Ixodes* is poorly represented, and very few *Hyalomma* and *Rhipicephalus* have been found in this region (Table 8.2). There are also relatively few representatives of the soft tick family Argasidae.

The tick faunal community can be defined at different levels: region, country, ecosystem, host, and microhabitat, and in the case of maritime countries the island

(s) occupied. Although recent species lists for mainland Southeast Asian countries are available in Petney et al. (2019), no such up-to-date lists are available for the island countries (note: *Dermacentor laothaiensis* and *D. pasteuri* were described after Petney et al. (2019) in Apanaskevich et al. 2019, 2020).

There are numerous country lists, most of which are dated. The older works listed in the introduction to this chapter have recently been supplemented by Phan (1977, Vietnam), Tanskul et al. (1983, Thailand), Kolonin (1995, Vietnam), and Kwak (2018, Singapore).

There are also some local or province lists that provide an indication of the tick faunas present at smaller scales. For example, Vongphayloth et al. (2016) surveyed the tick fauna of two protected areas in Nakai District, Khammouane Province, Lao PDR. Sampling was carried out by dragging vegetation in two habitats (mountainous primary forest and valley secondary forest), thus providing an indication of the tick species present in certain natural/semi-natural tick habitats within a defined political jurisdiction. Vongphayloth et al. (2016) identified 11 species belonging to five genera (*Amblyomma*, *Dermacentor*, *Haemaphysalis*, *Ixodes*, and *Rhipicephalus*).

A number of other local surveys, often focusing on a restricted host spectrum, have provided valuable information on broad-scale, potentially non-interacting tick species within tick communities (Hoogstraal et al. 1968, 1972; Durden and Watts 1989; Durden et al. 2008; Mariana et al. 2008; Ishak et al. 2018).

Because national boundaries are artificial constructs that generally do not conform to ecological zones, they are of relatively little value in defining the distributions of tick species. Of more relevance are the ecosystem characteristics themselves, as these define both the hosts found in an area and the microclimate available to ticks when they are detached from their host. This is of vital importance because appropriate temperatures and relative humidities are critical for tick survival (Needham and Teel 1991). Nevertheless, many tick species found in the same habitat may not be part of, or may exist peripheral to, an interactive community. This is often the case with birds, which may range over vast, ecologically unrelated areas. For example, *Haemaphysalis megalaimae* is a parasite specific to the avian family Megalaimidae, the Asian barbets (previously classified in the Capitonidae, the New World barbets), which are widespread in Southeast Asia (Guglielmone et al. 2014).

Although the tick fauna of Southeast Asia is relatively well known morphologically, considerable molecular work will be required to confirm some morphological descriptions. This is particularly true of several recently described *Dermacentor* species, all of which parasitize the ubiquitous wild boar (*Sus scrofa*) (Petney et al. 2019), calling the concept of allopatric speciation into question. The value of modern molecular methods can be gauged by the example of *Rhipicephalus* (*Boophilus*) *microplus* sensu lato (s.l.), which a decade ago was thought to constitute a single taxon distributed widely throughout the tropics. In 2012, Estrada-Peña et al. showed that the Australian population in fact represented a previously described species, *R.* (*B.*) *australis*. Thereafter, Roy et al. (2018) showed that *R.* (*B.*) *microplus* (s.l.) (Fig. 8.2) consists of at least five species and that morphological methods alone cannot effectively distinguish these. A similar situation exists in the case of the brown dog tick *Rhipicephalus sanguineus* (Fig. 8.3), which consists of at least two

**Fig. 8.2** *Rhipicephalus microplus* sensu lato females in different stages of engorgement from cattle in northeast Thailand (photo by Weerachai Saijuntha)

**Fig. 8.3** An engorged and an unengorged female *Rhipicephalus sanguineus* sensu lato from a dog in northeast Thailand (photo by Weerachai Saijuntha)

species (Nava et al. 2015; Sanches et al. 2016). As these species are frequently encountered and occur on common hosts, it seems likely that less abundant species with patchy distributions, particularly those found only on islands and that do not parasitize migratory birds, may well represent species complexes. The elucidation of such complexes is of obvious importance when considering the potential role of particular tick species in pathogen transmission.

## 8.3 Tick Genetic Diversity

Investigations of the genetic diversity of Southeast Asian ticks have largely focused on Malaysia, where the molecular phylogeny of *Ixodes granulatus* was described based on the mitochondrial CO1 sequence (Lah et al. 2014) and subsequently used to

**Fig. 8.4** *Ixodes kopsteini* female collected from a wrinkle-lipped free-tailed bat (*Chaerephon plicatus*, family Molossidae) from Phaya cave, Loei Province, Thailand (Photo by Weerachi Saijuntha)

identify the different life history stages of this tick (Lah et al. 2016). Additionally, the cattle tick *Rhipicephalus microplus* in Malaysia was characterized using 16S rRNA and CO1 sequences, revealing four genetically divergent groups within Malaysian *R. microplus* and the identification of three principal genetic assemblages worldwide (Low et al. 2015). More recently, CO1 sequence variation of Malaysian *Haemaphysalis* species has been investigated (Ernieenor et al. 2017). Of 19 immature tick specimens collected from four localities, 16 were molecularly identified as *Haemaphysalis hystricis*, and three as *H. humerosa*, with sequence homologies of 97–99% and 86–87%, respectively. Low intraspecific variation (<0.3%) but high interspecific value (>15%) among *H. hystricis* was also observed (Ernieenor et al. 2017).

Some investigations of Southeast Asian tick genetic diversity have been conducted in areas outside Malaysia, such as a study of Bm86 mRNA sequence variation among *R. microplus* populations collected from cattle across Thailand, which revealed the presence of 6–9 genetically distinct groups related to Thai geography, all clearly separable from Neotropical *R. microplus* (Kaewmongkol et al. 2015). In Vietnam, Hornok et al. (2015) presented molecular evidence to show that more than one species may exist under the names *Ixodes simplex* and *Ixodes vespertilionis*. And most recently, high genetic diversity has been reported in six species of hard ticks—*Haemaphysalis kitaokai*, *H. longicornis*, *H. shimoga*, *Ixodes ovatus*, *Rhipicephalus haemaphysaloides,* and *R. microplus*—from a China-Myanmar border county, where specimens were analyzed using 16S rRNA and CO1 sequences. New species or subspecies closely related to *H. kitaokai*, *H. shimoga*, *I. ovatus,* and *R. haemaphysaloides* probably exist in this area (Li et al. 2018).

## 8.4 Tick Biodiversity on Host Species

Individual hosts or host populations commonly harbor a number of tick species. Intensive studies of such on-host communities have been conducted throughout southern Africa, where tick-borne diseases of livestock are a major veterinary problem (MacLeod 1970; Horak et al. 1979, 2001; Smith and Parker 2010; Nyangiwe et al. 2011). The veterinary concerns caused by ticks and their economic impact led directly to investigations of southern African ticks and tick-borne diseases (De Kock 1945; Horak et al. 2018). Communities of different tick species also occur on wildlife and companion animals (Petney and Horak 1997; Neves et al. 2004; Matthee et al. 2010; Anderson et al. 2013). Such communities can influence the potential transmission dynamics of pathogens when more than one species can act as a vector for the same disease-causing agent (Pfäffle et al. 2013). If two or more species that transmit different pathogens occur on a host, then infection with multiple pathogens can occur, leading to greater morbidity or mortality (Petney and Andrews 1998).

Perhaps due to a lack of perceived medical and veterinary significance of ticks in Southeast Asia, relatively little research has been carried out on tick communities in this area. Tanskul et al. (1983) provide a tick-host list for Thailand, unfortunately without listing the number of hosts examined. Of the 39 bird species listed, a single tick species was found on 33, and 2 tick species were found on 6 bird species. Eight tick species were found on domestic cattle (*Bos* spp.), while 8 wild artiodactyl species were infested with an average of 3.6 tick species. Wild carnivores were found with an average of 2.6 tick species, insectivores with 3.0 (the common tree shrew, *Tupaia glis*, had 7 species), bats with an average of 1 species, lagomorphs with 3, tapir (Perissodactyla) with 4, pangolins with 1, and rodents with 2.3. Reptiles averaged 1.7 tick species/host species.

Such figures are all likely to be underestimates due to generally low host sample sizes and recent additions to the number of tick species known to be present. For example, humans, who are not natural hosts for ticks, were found to be parasitized by 15 tick species. Tanskul et al. (1983) list seven species from dogs in Thailand, whereas for mainland Southeast Asia 18 tick species have been recorded (Petney et al. 2019). Even given these caveats, the data clearly show that individual host species harbor tick communities with potential interactions between species.

There are numerous small-scale surveys, also often with low sample sizes, that nonetheless provide an indication of the potentially interactive tick community on individual host species. Robbins et al. (1997) identified three species of tick from an Asian golden cat (*Catopuma temminckii*), including adults of *Haemaphysalis asiatica* and *Rhipicephalus haemaphysaloides*, as well as nymphs of an undetermined *Amblyomma* species. Grassman Jr et al. (2004) collected six species of ticks from eight species of carnivore, finding four tick species on *Prionailurus bengalensis* (leopard cat), three on *Martes flavigula* (yellow-throated marten), *Neofelis nebulosa* (clouded leopard) and *Cuon alpinus* (dhole), two on *Catopuma temminckii*, and a single species on *Pardofelis marmorata* (marbled cat), *Arctictis binturong* (binturong), and *Viverra zibetha* (large Indian civet). Other Southeast Asian studies of ticks on wildlife include Audy et al. (1960), Munaf (1978), and Madinah et al. (2011).

Sahara et al. (2019) identified 1575 ticks on 26 cattle from five different areas of Indonesia. They found infestations with *Rhipicephalus microplus*, *Haemaphysalis bispinosa*, and *Rhipicephalus pilans*, with the cattle from three areas infested by both *R. microplus* and *R. pilans*. A survey within the Myanmar-Chinese border area of Yunnan Province (which is biogeographically related to mainland Southeast Asia, Yang et al. 2004, Zhu 2017) revealed that goats harbored four tick species (*R. microplus*, *R. haemaphysaloides*, *I. ovatus,* and *H. longicornis*) and cattle two (*R. microplus* and *R. haemaphysaloides*). The taxonomic situation was complicated because *R. haemaphysaloides* and *I. ovatus* each fell into three phylogenetic groups, with intergroup genetic distances higher than is normal between tick species. *Haemaphysalis longicornis* ticks also clustered into two separate clades, indicating that more than one species group could be present (Li et al. 2018).

Movement of hosts can also lead to the simultaneous transmission of one or more tick species. For example, a king cobra (*Ophiophagus hannah*) exported from Malaysia to Taiwan was infested with *Amblyomma cordiferum* and *A. varanense*, with females of both species being present (Norval et al. 2009). And the dramatic increase in commerce and communication between Southeast Asia and the rest of the world is almost certain to result in the importation of exotic tick species, as in the case of two male specimens of the Neotropical sloth tick *Amblyomma varium* collected from a male and female of the southern two-toed sloth (*Choloepus didactylus*) that were shipped to Wildlife Reserves Singapore in 2012 (both tick specimens, identified by RGR, have been deposited in the permanent research collections of the Peabody Museum of Natural History at Yale University, accession numbers YPM-ENT 300786–300787).

Very few tick species are specific parasites of a single host species. Most species have some preference for a group of hosts. Thus, a number of *Dermacentor* species occur most commonly on wild boar (Vongphayloth et al. 2018), while others prefer bats, tortoises, snakes and lizards, and particular mammal species (e.g., *Amblyomma javanense* on pangolins, *Manis* spp.) (Kwak et al. 2018; Petney et al. 2019). In addition, the immature stages of some tick species may have host preferences that differ from those of the adults (Petney et al. 2019). This mixture of hosts and life history stages complicates efforts to define tick communities (Petney and Horak 1997), requiring a knowledge of host species utilization by both adults and immatures in order to ascertain potential pathogen transmission pathways. Bats (Chiroptera) are unusual as they can act as hosts for both argasid and ixodid ticks, potentially allowing for the transfer of pathogens between these two families (Table 8.4).

**Table 8.4** Argasid and ixodid tick species recorded from bats in Southeast Asia. *Ixodes simplex* and *I. vespertilionis* may be species complexes (Guglielmone et al. 2020). *Ixodes kopsteini*, an unusual species in which the eggs hatch in the dead body of the female, is shown in Fig. 8.4

| Family | Tick species |
|---|---|
| Argasidae | *Argas pusillus* |
| | *Argas vespertilionis* |
| | *Ornithodoros batuensis* |
| Ixodidae | *Ixodes kopsteini* |
| | *Ixodes simplex* |
| | *Ixodes vespertilionis* |

## 8.5 Ticks as Vectors

Ticks are known to be major vectors of human and animal viral, bacterial, and protozoan diseases worldwide (de la Fuente et al. 2008; Guglielmone and Robbins 2018). Although this is also the case for Southeast Asia, comparatively little large-scale work has been done in this region (Petney et al. 2019). However, with the advent of molecular taxonomic methods, local studies have revealed a large number of hitherto unknown pathogenic agents, particularly members of the Rickettsiaceae (*Rickettsia*, *Ehrlichia,* and *Anaplasma*) (Petney et al. 2019; Low et al. 2020), the pathogenic potentials of which remain to be determined.

Petney et al. (2019) list a single argasid (20% of the known total number of Southeast Asian species) and 25 ixodids (26%) as potential vectors for human pathogens: one *Argas* species, three *Amblyomma*, five *Dermacentor*, one *Ixodes*, 13 *Haemaphysalis,* and three *Rhipicephalus*. This contrasts with potential vectors of pathogens of wildlife, stock and companion animals, with three (60%) argasids, all belonging to the genus *Argas*, and 14 ixodids (14%) (three *Amblyomma*, two *Dermacentor*, seven *Haemaphysalis,* and two *Rhipicephalus*). In both cases, a large number of unidentified tick species were associated with potential pathogens. Low et al. (2020) list 14 species of ixodid tick associated with rickettsiae in Southeast Asia (we know of no record of *Amblyomma integrum* from this region) from a variety of wild, companion and stock animals as well as humans.

## 8.6 Pathogen Identification

The advent of molecular methods has made pathogen identification relatively rapid and precise (Solano-Gallego et al. 2016; Lempereur et al. 2017; Seesao et al. 2017). It has, however, led to the discovery of a large number of pathogen species that have yet to be taxonomically defined, i.e., agents lacking a binomial name. Thus, in continental Southeast Asia, Petney et al. (2019) listed 14 known bacterial pathogens or potential pathogens of humans and 17 species with no biologically correct designation. For vertebrate animals the situation was slightly better for bacteria, with 16 known and 16 unnamed species, while there were eight species in each category for protozoa. Most of the publications commenting on unnamed microbial species are very recent, and the ticks or hosts in which the potential pathogens were discovered represent a minuscule subset of those that occur in the region. We therefore expect a large increase in the number of undescribed pathogen species as more tick species and hosts are examined.

## 8.7 Pathogen Diversity

Wild animals are a potential source of a variety of bacterial and protozoan pathogens in Southeast Asia. Kho et al. (2015) examined 12 *Amblyomma varanense* and nine *Amblyomma helvolum* from seven *Python molurus* (Indian rock python) and six *A. helvolum* from a single *Naja sumatrana* (equatorial spitting cobra) from Johore, Malaysia. They found two potentially novel spotted fever group rickettsiae in the ticks: *Candidatus* Rickettsia sepangensis was determined from an engorged *A. varanense* with a high sequence similarity to *Rickettsia tamurae. Candidatus* Rickettsia johorensis was present in two samples from *A. helvolum* and two *A. varanense* ticks; it is closely related to *Rickettsia raoultii. Anaplasma* and *Ehrlichia* DNA was also found in seven and two ticks, respectively. Thus, four potentially new pathogens were found in a small sample of ticks from two host snake species.

In a similar study, Sumrandee et al. (2014b) examined five *A. helvolum* from single specimens of *Python bivittatus bivittatus* (Burmese python), *Xenochrophis piscator* (Asiatic water snake), and *Ptyas korros* (Indo-Chinese rat snake) and found that all ticks contained rickettsia, while an additional four *Ophiophagus hannah* (king cobra) infested with 19 ticks were again all infected with rickettsia. Five potentially novel species of rickettsia were found, with some isolates sequencing close to *Rickettsia belli* and *R. raoultii*. In addition, ticks from the same snakes yielded *Francisella*-like species (Sumrandee et al. 2014a) and *Hepatozoon* species (Sumrandee et al. 2015). In their 2015 study, Sumrandee et al. found *Theileria* species in *Haemaphysalis lagrangei, Ha. obesa*, and *Rhipicephalus microplus* from sambar deer (*Rusa unicolor*) and *Hepatozoon* species in *Dermacentor auratus* and *D. atrosignatus* from wild boar.

Domestic and stray dogs also harbor a number of tick-borne pathogens throughout Southeast Asia, including *Babesia vogeli, B. canis, Ehrlichia canis*, and *Hepatozoon canis*, with multiple species infections possible (Irwin and Jefferies 2004; Inpankaew et al. 2016; Petney et al. 2019). The vectors potentially include 18 species of ticks that have been found infesting dogs (Petney et al. 2019).

There is also considerable scope for potential human infection. Fever of unknown origin is commonplace in Southeast Asia, and some infections are probably caused by tick-borne pathogens (Aung et al. 2014; Kho et al. 2016).

Parola et al. (2003a) tested ticks collected from animals, people, and vegetation on the Thai-Myanmar border and from Vietnam. Six hundred and fifty specimens representing 13 species were collected and analyzed for *Anaplasma*, *Ehrlichia,* and *Rickettsia* species. Three species each of *Anaplasma* and *Ehrlichia* were found, as well as two species of *Rickettsia*. They then examined the blood of 46 patients with fever for rickettsial infection and found that eight were infected with spotted fever group rickettsia, of which only one, an infection with *R. felis*, does not have a tick-host cycle (Parola et al. 2003b).

## 8.8 Dynamic Changes in Tick and Tick-Borne Pathogen Communities

Southeast Asia is undergoing a period of major change, including massive increases in the human population (Jones 2013), changes in land-use away from natural habitats to urban, suburban, and agricultural land (Zhao et al. 2006), and climate change (Petney et al. this volume). During the nineteenth century, European travelers, such as Mouhot (1863, 1864) Wallace (1869) and Warington-Smyth (1895), described Southeast Asian countries as having limited access to rural areas, with natural vegetation predominating over agricultural land. The same situation was reported in the 1930s by May (1949). Now, much of Southeast Asia is connected by roads, aviation, and marine and freshwater navigation (Bowen Jr 2016; Kaffashi et al. 2016). The transition from natural habitats to human-influenced environments has a number of implications for the tick fauna of this region:

1. Reduction in natural habitats will reduce the number of potential tick host species, particularly the larger species that require larger territories to survive. In many areas, this will reduce and potentially eliminate hosts for ticks with a limited host range.
2. Habitat fragmentation will have the same effect unless large enough areas are preserved (Lynam and Billick 1999; Trisurat et al. 2010; Gibson et al. 2013).
3. Removal of natural habitats will bring humans into potential contact with tick species transmitting novel pathogens (Estrada-Peña et al. 2014; Loh et al. 2016).
4. Increases in urban and peri-urban environments will lead to increases in synanthropic vertebrate species that will act as hosts for their own tick faunas (Klimant et al. 2015; Hassell et al. 2017; Morand et al. 2019).
5. A spectrum of tick-borne diseases will emerge in these altered environments (Rizzoli et al. 2014; Noden et al. 2017).
6. Increases in grazing and pasture land for stock animals will lead to increased populations of those tick species that feed on these hosts, with consequent increases in tick-transmitted diseases.
7. There will be a consequent increased need for tick control and for greater knowledge of tick-borne diseases by physicians and veterinarians.

## 8.9 Conclusions

Ticks are common parasites of wild, stock, and companion animals in Southeast Asia, where there is an increasing recognition of their medical and veterinary significance. This has led to a number of studies that show the complexity of the tick communities inhabiting different ecosystems and infesting different hosts, as well as to recognition of the potential pathogens that they do or could transmit. Although the tick fauna is relatively well known, little information is available concerning interactions within tick communities and their relation to the epidemiology of tick-borne diseases.

**Acknowledgments** The scientific and common names of all vertebrates discussed in this chapter were verified via the Integrated Taxonomic Information System (itis.gov), a partnership of U.S. federal agencies "formed to satisfy their mutual needs for scientifically credible taxonomic information." All material in this chapter has been reviewed by the Walter Reed Army Institute of Research. There is no objection to its presentation and/or publication. The opinions or assertions contained herein are the private views of the authors and are not to be construed as official or reflecting the true views of the U.S. Department of the Army or the Department of Defense.

## References

Ahantarig A, Trinachartvanit W, Milne JR (2008) Tick-borne pathogens and diseases of animals and humans in Thailand. Southeast Asian J Trop Med Public Health 39(6):1015–1032

Anastos G (1950) The scutate ticks, or Ixodidae, of Indonesia. Entomol Am 30(1):1–144

Anderson K, Ezenwa VO, Jolles AE (2013) Tick infestation patterns in free ranging African buffalo (*Syncercus caffer*): effects of host innate immunity and niche segregation among tick species. Int J Parasitol Parasites Wildl 2:1–9

Apanaskevich DA, Chaloemthanetphong A, Vongphayloth K, Ahantarig A, Apanaskevich MA, Brey PT, Hertz JC, Lakeomany K, Sutherland IW, Trinachartvanit W (2019) Description of a new species of *Dermacentor* Koch, 1844 (Acari: Ixodidae) from Laos and Thailand. Syst Parasitol 96(6):475–484

Apanaskevich DA, Vongphayloth K, Jeangkhwoa P, Chaloemthanetphong A, Ahantarig A, Apanaskevich MA, Brey PT, Lakeomany K, Trinachartvanit W (2020) Description of a new species of *Dermacentor* Koch, 1844 (Acari: Ixodidae) from the mountains of Laos and Thailand. Syst Parasitol 97:347–355

Audy JR, Nadchatram M, Lim BL (1960) Malaysian parasites XLIX. Host distribution of Malayan ticks (Ixodoidea). Stud Inst Med Res Fed Malay States 29:225–246

Aung AK, Spelman DW, Murray RJ, Graves S (2014) Rickettsial infections in Southeast Asia: implications for local populace and febrile returned travelers. Am J Trop Med Hyg 91 (3):451–460

Bickford D, Poo SI, Posa MR (2012) Southeast Asian biodiversity crisis. In: Gower D, Johnson K, Richardson J, Rosen B, Rüber L, Williams S (eds) Biotic evolution and environmental change in Southeast Asia. Cambridge University Press, Cambridge, pp 434–462

Bordes F, Morand S, Pilosof S, Claude J, Krasnov BR, Cosson JF, Chaval Y, Ribas A, Chaisiri K, Blasdell K, Herbreteau V (2015) Habitat fragmentation alters the properties of a host–parasite network: rodents and their helminths in South-East Asia. J Anim Ecol 84(5):1253–1263

Bowen JT Jr (2016) "Now everyone can fly"? Scheduled airline services to secondary cities in Southeast Asia. J Air Transp Manag 53:94–104

Ciofi C, Beaumont MA, Swingland IR, Bruford MW (1999) Genetic divergence and units for conservation in the Komodo dragon *Varanus komodoensis*. Proc R Soc Lond Ser B Biol Sci 266 (1435):2269–2274
Crooks KR, Burdett CL, Theobald DM, King SR, Di Marco M, Rondinini C, Boitani L (2017) Quantification of habitat fragmentation reveals extinction risk in terrestrial mammals. Proc Natl Acad Sci U S A 114(29):7635–7640
De Kock G (1945) Tick infestation as an aetiological factor in diseases of unknown origin among cattle. J S Afr Vet Assoc 16(1):65–69
de la Fuente J, Estrada-Pena A, Venzal JM, Kocan KM, Sonenshine DE (2008) Overview: ticks as vectors of pathogens that cause disease in humans and animals. Front Biosci 13(13):6938–6946
Durden LA, Keirans JE (1996) Host–parasite coextinction and the plight of tick conservation. Am Entomol 42(2):87–91
Durden LA, Watts CHS (1989) A collection of ticks (Ixodidae) from Sulawesi Utara, Indonesia. Southeast Asian J Trop Biol 2:32–37
Durden LA, Merker S, Beati L (2008) The tick fauna of Sulawesi, Indonesia (Acari: Ixodoidea: Argasidae and Ixodidae). Exp Appl Acarol 45(1–2):85–110
Ernieenor FCL, Ernna G, Mariana A (2017) Phenotypic and genotypic identification of hard ticks of the genus *Haemaphysalis* (Acari: Ixodidae) in Peninsular Malaysia. Exp Appl Acarol 71 (4):387–400
Estrada-Peña A, Venzal JM, Nava S, Mangold A, Guglielmone AA, Labruna MB, De La Fuente J (2012) Reinstatement of *Rhipicephalus* (*Boophilus*) *australis* (Acari: Ixodidae) with redescription of the adult and larval stages. J Med Entomol 49(4):794–802
Estrada-Peña A, Ostfeld RS, Peterson AT, Poulin R, de la Fuente J (2014) Effects of environmental change on zoonotic disease risk: an ecological primer. Trends Parasitol 30(4):205–214
Gatti RC, Liang J, Velichevskaya A, Zhou M (2019) Sustainable palm oil may not be so sustainable. Sci Total Environ 652:48–51
Gibson L, Lynam AJ, Bradshaw CJ, He F, Bickford DP, Woodruff DS, Bumrungsri S, Laurance WF (2013) Near-complete extinction of native small mammal fauna 25 years after forest fragmentation. Science 341(6153):1508–1510
Grassman LI Jr, Sarataphan N, Tewes ME, Silvy NJ, Nakanakrat T (2004) Ticks (Acari: Ixodidae) parasitizing wild carnivores in Phu Khieo wildlife sanctuary, Thailand. J Parasitol 90 (3):657–659
Gray TN, Hughes AC, Laurance WF, Long B, Lynam AJ, O'Kelly H, Ripple WJ, Seng T, Scotson L, Wilkinson NM (2017) The wildlife snaring crisis: an insidious and pervasive threat to biodiversity in Southeast Asia. Biodivers Conserv 27(4):1031–1037
Guglielmone AA, Robbins RG (2018) Hard ticks (Acari: Ixodida: Ixodidae) parasitizing humans. A global overview. Springer, Heidelberg
Guglielmone AA, Robbins RG, Apanaskevich DA, Petney TN, Estrada-Peña A, Horak IG (2014) The hard ticks of the world. Springer, New York
Guglielmone AA, Petney TN, Robbins RG (2020) Ixodidae (Acari: Ixodoidea): descriptions and redescriptions of all known species from 1758 to December 31, 2019. Zootaxa 4871(1):1–322
Hassell JM, Begon M, Ward MJ, Fèvre EM (2017) Urbanization and disease emergence: dynamics at the wildlife–livestock–human interface. Trends Ecol Evol 32(1):55–67
Hoogstraal H, Santana FJ, Van Peenen PD (1968) Ticks (Ixodoidea) of Mt. Sontra, Danang, Republic of Vietnam. Ann Entomol Soc Am 61(3):722–729
Hoogstraal H, Lim BL, Nadchatram M, Anastos G (1972) The Gunong Benom expedition 1967. 8. Ticks (Ixodidae) of Gunong Benom and their altitudinal distribution, hosts and medical relationships. Bull Brit Mus (Nat Hist) D Zool 23(7):165–186
Horak IG, Londt JG, De Villiers IL (1979) Parasites of domestic and wild animals in South Africa. XIII. The seasonal incidence of adult ticks (Acarina: Ixodidae) on cattle in the northern Transvaal. Onderstepoort J Vet Res 46(1):31–39

Horak IG, Emslie FR, Spickett AM (2001) Parasites of domestic and wild animals in South Africa. XL. Ticks on dogs belonging to people in rural communities and carnivore ticks on the vegetation. Onderstepoort J Vet Res 68(2):135–141

Horak IG, Heyne H, Williams R, Gallivan GJ, Spickett AM, Bezuidenhout JD, Estrada-Peña A (2018) The ixodid ticks (Acari: Ixodidae) of southern Africa. Springer, Heidelberg

Hornok S, Estrada-Peña A, Kontschán J, Plantard O, Kunz B, Mihalca AD, Thabah A, Tomanović S, Burazerović J, Takás N, Görföl T, Estók P, Tu VT, Szőke K, de Mera IGF, de la Fuente J, Takahashi M, Yamauchi T, Takano A (2015) High degree of mitochondrial gene heterogeneity in the bat tick species *Ixodes vespertilionis*, *I. ariadnae* and *I. simplex* from Eurasia. Parasit Vectors 8:457

Hughes AC (2017) Understanding the drivers of Southeast Asian biodiversity loss. Ecosphere 8 (1):1–33

Imai N, Furukawa T, Tsujino R, Kitamura S, Yumoto T (2018) Factors affecting forest area change in Southeast Asia during 1980-2010. PLoS One 13(5):e0197391

Inpankaew T, Hii SF, Chimnoi W, Traub RJ (2016) Canine vector-borne pathogens in semi-domesticated dogs residing in northern Cambodia. Parasit Vectors 9(1):253

Irwin PJ, Jefferies R (2004) Arthropod-transmitted diseases of companion animals in Southeast Asia. Trends Parasitol 20(1):27–34

Ishak SN, Yusof MA, Nor SM, Sah SA, Lim FS, Khoo JJ, Mohd-Taib FS (2018) Prevalence of on-host ticks (Acari: Ixodidae) in small mammals collected from forest near to human vicinity in Selangor, Malaysia. Syst Appl Acarol 3(8):1531–1544

Jones GW (2013) The population of Southeast Asia. Asia research institute working paper series no. 196. National University of Singapore, Singapore, pp 1–39. https://themimu.info/sites/themimu.info/files/documents/Ref%20Doc_The%20Population%20of%20SE%20Asia_ARI_Jan13.pdf

Kaewmongkol S, Kaewmongkol G, Inthong N, Lakkitjaroen N, Sirinarumitr T, Berry CM, Jonsson NN, Stich RW, Jittapalapong S (2015) Variation among Bm86 sequences in *Rhipicephalus* (*Boophilus*) *microplus* ticks collected from cattle across Thailand. Exp Appl Acarol 66 (2):247–256

Kaffashi S, Shamsudin MN, Clark MS, Sidique SF, Radam A (2016) Road transport system in Southeast Asia; problems and economic solutions. Curr World Environ 11(1):1–15

Kho KL, Koh FX, Tay ST (2015) Molecular evidence of potential novel spotted fever group rickettsiae, *Anaplasma* and *Ehrlichia* species in *Amblyomma* ticks parasitizing wild snakes. Parasit Vectors 8(1):1–5

Kho KL, Koh FX, Singh HK, Zan HA, Kukreja A, Ponnampalavanar S, Tay ST (2016) Spotted fever group rickettsioses and murine typhus in a Malaysian teaching hospital. Am J Trop Med Hyg 95(4):765–768

Klimant P, Baláž I, Krumpálová Z (2015) Communities of small mammals (Soricomorpha, Rodentia) in urbanized environment. Biologia 70(6):839–845

Kohls GM (1950) Ticks (Ixodoidea) of the Philippines. Natl Inst Health Bull 192:1–28

Kohls GM (1957) Malaysian parasites-XVIII ticks (Ixodoidea) of Borneo and Malaya. Stud Inst Med Res Malaya 28:65–94

Kolonin GV (1995) Review of the ixodid tick fauna (Acari: Ixodidae) of Vietnam. J Med Entomol 32(3):276–282

Kwak ML (2018) Ticks in the Lion City: a preliminary review of the tick fauna of Singapore. Exp Appl Acarol 76(2):263–267

Kwak ML, Hsu CD, Douay G, Ahmad AA (2018) The first authenticated record of the pangolin tick *Amblyomma javanense* (Acari: Ixodidae) in Singapore, with notes on its biology and conservation. Exp Appl Acarol 76(4):551–557

Lah EFC, Yaakop S, Ahamad M, George E, Nor SM (2014) Molecular phylogeny of a tick, *Ixodes granulatus* (Acari: Ixodidae) based on cytochrome oxidase subunit I (COI) marker. AIP Conf Proc 1614:719–726

Lah EFC, Yaakop S, Ahamad M, George E, Nor SM (2016) Precise identification of different stages of a tick, *Ixodes granulatus* Supino, 1897 (Acari: Ixodidae). Asian Pac J Trop Biomed 6 (7):597–604

Lempereur L, Beck R, Fonseca I, Marques C, Duarte A, Santos M, Zúquete S, Gomes J, Walder G, Domingos A, Antunes S (2017) Guidelines for the detection of *Babesia* and *Theileria* parasites. Vector Borne Zoonotic Dis 17(1):51–65

Li LH, Zhang Y, Wang JZ, Li XS, Yin SQ, Zhu D, Xue JB, Li SG (2018) High genetic diversity in hard ticks from a China-Myanmar border county. Parasit Vectors 11(1):469

Loh E, Murray KA, Nava A, Aguirre A, Daszak P (2016) Evaluating the links between biodiversity, land-use change, and infectious disease emergence in tropical fragmented landscapes. Trop Conserv Perspect Local Glob Prior 13:79

Low VL, Tay ST, Kho KL, Koh FX, Tan TK, Lim YAL, Ong BL, Panchadcharam C, Norma-Rashid Y, Sofian-Azirun M (2015) Molecular characterisation of the tick *Rhipicephalus microplus* in Malaysia: new insights into the cryptic diversity and distinct genetic assemblages throughout the world. Parasit Vectors 8:341

Low VL, Tan TK, Khoo JJ, Lim FS, AbuBakar S (2020) An overview of rickettsiae in Southeast Asia: vector-animal-human interface. Acta Trop 202:105282

Lynam AJ, Billick I (1999) Differential responses of small mammals to fragmentation in a Thailand tropical forest. Biol Conserv 91(2–3):191–200

MacLeod J (1970) Tick infestation patterns in the southern province of Zambia. Bull Entomol Res 60(2):253–274

Madinah A, Fatimah A, Mariana A, Abdullah MT (2011) Ectoparasites of small mammals in four localities of wildlife reserves in peninsular Malaysia. Southeast Asian J Trop Med Publ Health 42(4):803–813

Mariana A, Zuraidawati Z, Ho TM, Kulaimi BM, Saleh I, Shukor MN, Shahrul-Anuar MS (2008) Ticks (Ixodidae) and other ectoparasites in Ulu Muda Forest Reserve, Kedah, Malaysia. Southeast Asian J Trop Med Public Health 39(3):496–506

Matthee S, Lovely C, Gaugler A, Beeker R, Venter HR, Horak IG (2010) Ixodid ticks on domestic dogs in the northern Cape Province of South Africa and in Namibia. J S Afr Vet Assoc 81 (2):126–128

May JM (1949) Siam doctor. Exotic adventures, medical and human, in the Far East. Doubleday, New York

Mihalca AD, Gherman CM, Cozma V (2011) Coendangered hard-ticks: threatened or threatening? Parasit Vectors 4(1):71

Morand S, Blasdell K, Bordes F, Buchy P, Carcy B, Chaisiri K, Chaval Y, Claude J, Cosson JF, Desquesnes M, Jittapalapong S (2019) Changing landscapes of Southeast Asia and rodent-borne diseases: decreased diversity but increased transmission risks. Ecol Appl 29(4):e01886

Mouhot H (1863, 1864) Travels in the central parts of Indo-China, Cambodia and Laos during the years 1858, 1859, and 1860. John Murray, London

Munaf HB (1978) Tick fauna of Baluran Wildlife Reserve, Indonesia. Hemera Zoa 70(1):37–44

Nava S, Estrada-Peña A, Petney T, Beati L, Labruna MB, Szabó MP, Venzal JM, Mastropaolo M, Mangold AJ, Guglielmone AA (2015) The taxonomic status of *Rhipicephalus sanguineus* (Latreille, 1806). Vet Parasitol 208(1–2):2–8

Needham GR, Teel PD (1991) Off-host physiological ecology of ixodid ticks. Annu Rev Entomol 36(1):659–681

Neves L, Afonso S, Horak IG (2004) Ixodid ticks on dogs in southern Mozambique. Onderstepoort J Vet Res 71(4):279–283

Noden BH, Loss SR, Maichak C, Williams F (2017) Risk of encountering ticks and tick-borne pathogens in a rapidly growing metropolitan area in the US Great Plains. Ticks Tick Borne Dis 8 (1):119–124

Norval G, Robbins RG, Kolonin G, Shiao PJ, Mao JJ (2009) Unintentional transport of ticks into Taiwan on a king cobra (*Ophiophagus hannah*). Herpetol Notes 2:203–206

Nyangiwe N, Goni S, Hervé-Claude LP, Ruddat I, Horak IG (2011) Ticks on pastures and on two breeds of cattle in the eastern Cape Province, South Africa. Onderstepoort J Vet Res 78(1):1–9

Parola P, Cornet JP, Sanogo YO, Miller RS, Van Thien H, Gonzalez JP, Raoult D, Telford SR III, Wongsrichanalai C (2003a) Detection of *Ehrlichia* spp., *Anaplasma* spp., *Rickettsia* spp., and other eubacteria in ticks from the Thai-Myanmar border and Vietnam. J Clin Microbiol 41 (4):1600–1608

Parola P, Miller RS, McDaniel P, Telford IIISR, Rolain JM, Wongsrichanalai C, Raoult D (2003b) Emerging rickettsioses of the Thai-Myanmar border. Emerg Infect Dis 9(5):592–595

Petney TN, Andrews RH (1998) Multiparasite communities in animals and humans: frequency, structure and pathogenic significance. Int J Parasitol 28(3):377–393

Petney TN, Horak IG (1997) The community structure of ticks on kudu, *Tragelaphus strepsiceros*, in the eastern Cape Province of South Africa. In: Dettner K, Bauer G, Völkl W (eds) Vertical food web interactions: evolutionary patterns and driving forces. Springer, Heidelberg, pp 81–89

Petney TN, Kolonin GV, Robbins R (2007) Southeast Asian ticks (Acari: Ixodida): an historical perspective. Parasitol Res 101(Suppl 2):201–205

Petney TN, Saijuntha W, Boulanger N, Chitimia-Dobler L, Pfeffer M, Eamudomkarn C, Andrews RH, Ahamad M, Putthasorn N, Muders SV, Petney DA, Robbins RG (2019) Ticks (Argasidae, Ixodidae) and tick-borne diseases of continental Southeast Asia. Zootaxa 4558(1):1–89

Pfäffle M, Littwin N, Muders SV, Petney TN (2013) The ecology of tick-borne diseases. Int J Parasitol 43(12–13):1059–1077

Phan TC (1977) Ve bet trung ky sinh Viet Nam. Tap Ve (Ixodoidea). Mo phan loai. Hanoi (in Vietnamese)

Rizzoli A, Silaghi C, Obiegala A, Rudolf I, Hubálek Z, Földvári G, Plantard O, Vayssier-Taussat M, Bonnet S, Špitalská E, Kazimírová M (2014) *Ixodes ricinus* and its transmitted pathogens in urban and peri-urban areas in Europe: new hazards and relevance for public health. Front Public Health 2:251

Robbins RG, Platt SG (2001) First report of *Amblyomma supinoi* Neumann (Acari: Ixodida: Ixodidae) from the Arakan Forest Turtle, *Geoemyda depressa* Anderson (Reptilia: Testudines: Emydidae), with additional records of this tick from the Union of Myanmar. Proc Entomol Soc Wash 103(4):1023–1024

Robbins RG, Karesh WB, Rosenberg S, Schonwalter N, Inthavong C (1997) Two noteworthy collections of ticks (Acari: Ixodida: Ixodidae) from endangered carnivores in the Lao People's Democratic Republic. Entomol News 108(1):60–62

Roy BC, Estrada-Peña A, Krücken J, Rehman A, Nijhof AM (2018) Morphological and phylogenetic analyses of *Rhipicephalus microplus* ticks from Bangladesh, Pakistan and Myanmar. Ticks Tick Borne Dis 9(5):1069–1079

Sahara A, Nugraheni YR, Patra G, Prastowo J, Priyowidodo D (2019) Ticks (Acari: Ixodidae) infestation on cattle in various regions in Indonesia. Vet World 12(11):1755–1759

Sanches GS, Évora PM, Mangold AJ, Jittapalapong S, Rodriguez-Mallon A, Guzmán PE, Bechara GH, Camargo-Mathias MI (2016) Molecular, biological, and morphometric comparisons between different geographical populations of *Rhipicephalus sanguineus* sensu lato (Acari: Ixodidae). Vet Parasitol 215:78–87

Schipper J, Chanson JS, Chiozza F et al (2008) The status of the world's land and marine mammals: diversity, threat, and knowledge. Science 322(5899):225–230

Seesao Y, Gay M, Merlin S, Viscogliosi E, Aliouat-Denis CM, Audebert C (2017) A review of methods for nematode identification. J Microbiol Methods 138:37–49

Sharifah N, Heo CC, Ehlers J, Houssaini J, Tappe D (2020) Ticks and tick-borne pathogens in animals and humans in the island nations of Southeast Asia: a review. Acta Trop 22:105527

Smith ER, Parker DM (2010) Tick communities at the expanding wildlife/cattle interface in the eastern Cape Province, South Africa: implications for corridor disease. J S Afr Vet Assoc 81 (4):237–240

Smith SB, Gotoh T, Greenwood PL (2018) Current situation and future prospects for global beef production: overview of special issue. Asian-Austral. J Anim Sci 31(7):927–932

Sodhi NS, Koh LP, Brook BW, Ng PK (2004) Southeast Asian biodiversity: an impending disaster. Trends Ecol Evol 19(12):654–660
Sodhi NS, Posa MRC, Lee TM et al (2010) The state and conservation of Southeast Asian biodiversity. Biodivers Conserv 19:317–328
Solano-Gallego L, Sainz Á, Roura X, Estrada-Peña A, Miró G (2016) A review of canine babesiosis: the European perspective. Parasit Vectors 9(1):336
Sumrandee C, Hirunkanokpun S, Grubhoffer L, Baimai V, Trinachartvanit W, Ahantarig A (2014a) Phylogenetic relationships of *Francisella*-like endosymbionts detected in two species of *Amblyomma* from snakes in Thailand. Ticks Tick Borne Dis 5(1):29–32
Sumrandee C, Hirunkanokpun S, Doornbos K, Kitthawee S, Baimai V, Grubhoffer L, Trinachartvanit W, Ahantarig A (2014b) Molecular detection of *Rickettsia* species in *Amblyomma* ticks collected from snakes in Thailand. Ticks Tick Borne Dis 5(6):632–640
Sumrandee C, Baimai V, Trinachartvanit W, Ahantarig A (2015) *Hepatozoon* and *Theileria* species detected in ticks collected from mammals and snakes in Thailand. Ticks Tick Borne Dis 6 (3):309–315
Tanskul P, Stark HE, Inlao I (1983) A checklist of ticks of Thailand (Acari: Metastigmata: Ixodoidea). J Med Entomol 20(3):330–341
Tölle MH, Engler S, Panitz HJ (2017) Impact of abrupt land cover changes by tropical deforestation on southeast Asian climate and agriculture. J Clim 30(7):2587–2600
Toumanoff C (1944) Les tiques (Ixodoidea) de l'Indochine. Saigon
Traub RJ, Irwin P, Dantas-Torres F, Tort GP, Labarthe NV, Inpankaew T et al (2015) Toward the formation of a Companion Animal Parasite Council for the Tropics (CAPCT). Parasit Vectors 8:271
Trisurat Y, Alkemade R, Verburg PH (2010) Projecting land-use change and its consequences for biodiversity in northern Thailand. Environ Manag 45(3):626–639
Vongphayloth K, Brey PT, Robbins RG, Sutherland IW (2016) First survey of the hard tick (Acari: Ixodidae) fauna of Nakai District, Khammouane Province, Laos, and an updated checklist of the ticks of Laos. Syst Appl Acarol 21(2):166–180
Vongphayloth K, Hertz JC, Lakeomany K, Apanaskevich DA, Robbins RG, Sutherland IW, Brey PT (2018) The genus *Dermacentor* (Acari: Ixodidae) in Laos: a review and update of species records. J Med Entomol 55(4):1047–1050
Wallace AR (1869) The Malay archipelago: the land of the orang-utan and the bird of paradise. A narrative of travel with studies of man and nature. 2 volumes. Macmillan, London
Warington-Smyth H (1895) Notes on a journey on the upper Mekong, Siam. R Geogr Soc, London
Yaap B, Struebig MJ, Paoli G, Koh LP, Koh LP (2010) Mitigating the biodiversity impacts of oil palm development. CAB Rev 5:19
Yang Y, Tian K, Hao J, Pei S, Yang Y (2004) Biodiversity and biodiversity conservation in Yunnan, China. Biodivers Conserv 13(4):813–826
Zhao S, Peng C, Jiang H, Tian D, Lei X, Zhou X (2006) Land use change in Asia and the ecological consequences. Ecol Res 21(6):890–896
Zhu H (2017) A biogeographical study on tropical flora of southern China. Ecol Evol 7 (23):10398–10408
Ziegler AD, Fox JM, Xu J (2009) The rubber juggernaut. Science 324(5930):1024–1025

# Chapter 9
# Parasite Diversity, Dynamics, and Climate Change

**Trevor N. Petney, Paiboon Sithithaworn, and Ross H. Andrews**

**Abstract** Global climate change is now evident across the globe and appears to be accelerating. In Southeast Asia this will lead to substantial increases in temperature and regional changes in rainfall patterns with some areas experiencing wetter conditions and others increased drought. Complicating factors include severe weather events, sea level rise, and drought-induced wild fires. As parasites are reliant on the external environment during parts of their life cycle and are endothermic, such changes will influence both the chances of parasite survival and population dynamics. In addition, hosts will also be influenced by these factors compounding the effects of climate. Our database is still limited and studies of local and regional parasite communities are essential for our understanding of climate impact on human and animal health.

**Keywords** Climate change · Global warming · Parasite diversity · Southeast Asia

---

T. N. Petney (✉)
Departments of Zoology and Paleontology and Evolution, State Museum of Natural History, Karlsruhe, Germany

Cholangiocarcinoma Research Institute, Khon Kaen University, Khon Kaen, Thailand

P. Sithithaworn
Cholangiocarcinoma Research Institute, Faculty of Medicine, Khon Kaen University, Khon Kaen, Thailand

R. H. Andrews
Cholangiocarcinoma Research Institute, Faculty of Medicine, Khon Kaen University, Khon Kaen, Thailand

Faculty of Medicine, Imperial College, London, UK

 
T. N. Petney et al. (eds.), *Biodiversity of Southeast Asian Parasites and Vectors causing Human Disease*, Parasitology Research Monographs 14,
https://doi.org/10.1007/978-3-030-71161-0_9

## 9.1 Introduction

Global climate change does and will continue to have a major impact on the environments in which animals live (Nunez et al. 2019; Sheldon 2019; Román-Palacios and Wiens 2020), and in spite of a growing recognition of its general significance for the human population, insufficient is being done to reduce it to an "acceptable" level (IPCC 2019; Forster et al. 2020). Our knowledge of climate change is an ongoing process with ever improving predictive models providing a basis for our understanding of future scenarios (O'Neill et al. 2017; Zhu et al. 2019). A very recent estimate suggests that the most likely range of average temperature increase with doubling of the $CO_2$ concentration in the atmosphere lies between 2.6 °C and 3.9 °C with an increase of 4.5 °C being possible although unlikely (Sherwood et al. 2020). In addition to increasing temperatures and changing patterns of rainfall, severe weather events, sea level rise, and drought-induced wild fires will become more frequent, leading to additional environmental damage (Anticamara and Go 2017; Van de Pol et al. 2017; Salmo III et al. 2019; Lin et al. 2020; Stephens et al. 2020). These changes will have a major effect on the human population (Xu et al. 2020) and the parasites using humans and animals as hosts (Carlson et al. 2017).

With respect to parasite biodiversity, climate change will affect both the parasite in its free-living environment and the hosts which it requires to complete its life cycle. Potential effects relating to parasites include changes in the environment leading to reduced biodiversity, host extinctions or invasions in certain areas, increases or decreases in habitats for vectors or intermediate life history stages and changes in population dynamics, particularly for species whose breeding season is influenced by temperature and rainfall (Marcogliese 2008, 2016; Rohr et al. 2011; Chevalier et al. 2016; Short et al. 2017). Temperature related changes may include a speeded up life cycle leading to more generations/year, increasing the area of transmission, or pathogenicity, or possibly breaking the synchronization between host and parasite (Zhou et al. 2008; Mas-Coma et al. 2009; Polley and Thompson 2009; Paull and Johnson 2011; Short et al. 2017).

Rohr et al. (2011) predict that specialist parasites and those with complex life cycles are more likely to go extinct than generalist parasites and that range shifts could expose hosts to novel parasites, which might lead to more severe disease. In contrast, Molnár et al. (2013) suggest a "shelter effect" in which behavioral thermoregulation by the intermediate host can act as a buffer for the larvae of indirectly transmitted parasites against the temperature extremes found as a consequence of global warming. Clearly, before any conclusions can be drawn sufficient field data must be collected from a variety of parasite species before any generalization can be made.

In addition, drought, flooding, or increasing temperatures may lead to tipping points beyond which the parasite or a host cannot survive (Altizer et al. 2013; Cizauskas et al. 2017). These could be induced by metabolic ecology (how well environmentally buffered the parasite species is), host body size, with larger wild hosts being more susceptible to extinction, and the ecology of transmission and persistence. The ability to switch hosts can lead to the survival of a parasite species but also harbors the possibility of increasing pathogenicity (Rohr et al. 2011), something that we are all now aware of (Touati et al. 2020). Carlson et al. (2017) estimated extinction rates for eight major parasite clades. Their model suggests that 5–10% of species will become extinct by 2070 from habitat loss due to climate change. They found that ectoparasites such as ticks are more likely to become extinct than endoparasites. The extinction of hosts may lead to up to 30% of parasitic worm species dying out. Nevertheless, in some areas parasite species richness may increase due to climate driven invasions.

## 9.2 The Situation in Southeast Asia

Although models of climate change predict substantial changes in temperature (Zhu et al. 2020) and precipitation, local variation is likely to occur (Tangang et al. 2019). Thus, general models covering the entirety of Southeast Asia may not be locally valid. Nevertheless, Amnuaylojaroen and Chanvichit (2019) predict a general reduction in rainfall by (−1)–(1) mm/day, with temperature increasing by up to 2–3 °C. The authors predict a higher likelihood of drought in both the dry and rainy seasons. Based on projected rainfall changes, simulations predict a significant decrease in rainfall over western maritime Southeast Asia during the inter-monsoon periods but no change during the wet season (Kang et al. 2019). Ge et al. (2019) show, based on their model, that extreme precipitation events are likely to increase across Southeast Asia with temperature increases of 1.5 °C and 2.0 °C.

Tangang et al. (2019) modelled potential changes for rainfall in Thailand to the end of the twenty-first century based on regional changes in the summer and winter monsoons. They showed regional variation between the southern areas and the rest of the country with respectively up to 15% drier and wetter conditions during the dry season (October to May). During the wet season (June to September), drier conditions are predicted for the whole country with the magnitude of change being >40% in some cases. The current situation can be gauged by considering Fig. 9.1. Currently monsoon rainfall leads to temporary, massive increases in flow levels in rivers and streams with consequent inundation of flood plains. Either increasing or decreasing levels of rainfall would have significant impacts on local freshwater ecosystems.

For Malaysia, Ngai et al. (2020) predict a decrease in the rainfall frequency over Malaysia with rainfall intensity and extremes over the Peninsular Malaysia likely to decrease in winter and increase in summer and autumn towards 2100. An extreme decrease in rainfall frequency and extremes is predicted for western Peninsular Malaysia (−4 to −8%). In some areas the likelihood of drought will increase (Tang et al. 2019).

The island states of Indonesia and the Philippines are particularly susceptible to climate change. The Philippines is predicted to have higher water availability with increased but more variable patterns of river flow and flooding (Tolentino et al.

**Fig. 9.1** The Songkhram River in the northeast of Thailand. During the rainy season the water level can rise by as much as 18 m with a flood plain covering up to $1850km^2$ in an exceptional year (Petney et al. 2009) (photo T.N. Petney taken at the beginning of the dry season 2012)

2016; Cabrera and Lee 2018). Rainfall patterns in Indonesia are tied to the presence or absence of El Nino (Chandrasa and Montenegro 2020) which are exacerbated by climate change (Thirumalai et al. 2017). Based on models and observational data, El Niño substantial increased the probability decreasing rainfall in the dry season in Indonesia with increasing temperatures due to human-induced climate change (King et al. 2016). In general, the El Nino Southern Oscillation (ENSO) has a major effect on the Southeast Asian climate. It is also increasing in frequency (Freund et al. 2019) and will continue to do so even after the global mean temperature has stabilized (Wang et al. 2017). During the 2015–2016 El Nino event there was a significant drought leading to substantially reduced photosynthesis (Qian et al. 2019). In general it has a significant effect on the natural and agricultural environments (Ismail and Chan 2019; Qian et al. 2019). El Nino will reduce precipitation whereas La Nina will increase it (Sun et al. 2020). ENSO also interacts with human driven climate change leading to positively feedback on increasing temperature and the number of drought events (Thirumalai et al. 2017; Rifai et al. 2019).

As with precipitation, the likelihood of extreme heat waves with high temperatures and long duration (>60 days) will increase in Southeast Asia by the end of the century (Jia et al. 2019; Villafuerte et al. 2020). For Borneo, models indicate that the minimum temperature will increase by 3.3–4.7 °C and the maximum by 3.0–4.6 °C.

This is a very substantial increase which will significantly impact the ecology of this island (Sa'adi et al. 2020). In particular, the increase in night temperatures is likely to be higher than during the daytime (Li 2020; Sa'adi et al. 2020).

With reduced precipitation and higher temperatures, there is also likely to be an increase in wild fires (Vadrevu et al. 2019). El Nino is also associated with an increasing risk of wildfires leading to a measurable increase in atmospheric CO2, thus exacerbating climate change.

In addition to changes in precipitation and temperature, other climate associated changes include sea level rise affecting low lying areas (in southern Vietnam there is also a significant subsidence of land particularly in the Mekong Delta area) (Erban et al. 2014; Jevrejeva et al. 2016; Tang 2019). For Southeast Asia the estimated mean sea level rise is 0.3 m by 2040 (Jevrejeva et al. 2016).

To summarize, although local variation is possible the general picture is for increasing temperatures, more frequent extreme weather events both in terms of heavy precipitation as well of droughts, wildfires and, in low lying, coastal areas, loss of habitat due to increasing sea levels.

## 9.3 Influence on Parasite Biodiversity

Climate change can influence parasite biodiversity in a number of ways via a number of causes depending on local conditions. Causal factors leading to biodiversity changes are higher temperatures, more or less rainfall, extreme weather events, and rising sea levels. The biomes present are mangroves, tropical forests (dry and moist broadleaf) and freshwater ecosystems including lakes and rivers (Olson et al. Olson and Dinerstein 2002; Abell et al. 2008). In addition, some parasite species and their hosts can be found plentifully in agricultural settings, for example in commonly found rice paddy-fish systems and aquaculture (dos Santos and Howgate 2011; Kiatsopit et al. 2012; Petney et al. 2013; Clausen et al. 2015) as well as in urban areas (Anh et al. 2007; Zain et al. 2012; Himsworth et al. 2013; Simonsen and Mwakitalu 2013).

Significant changes in climate, such as drought, frequent flooding and wildfires, or indeed sea-level rise, can cause sufficient changes to the physico-chemical environment to lead to niche modification and a reorganization of the animal and plant community (Walther et al. 2002; Sheldon 2019), including the vectors and hosts of parasites (Brooks and Hoberg 2007; Cizauskas et al. 2017). Thus, consideration of climate change and parasites implies changes in host and parasite biodiversity.

## 9.4 Mangroves

Changes to mangroves along the coastline of Southeast Asia will predominantly be due to rising sea levels (Saintilan et al. 2020). These authors found that there is a >90% likelihood that mangroves will not adapt to a sea level rise of >6.1 mm/year; however, the sea level rise is predicted to reach more than 7 mm/year by 2050 (Church et al. 2013; Bamber et al. 2019).

From the parasite point of view, mosquitoes and mosquito transmitted parasites are most likely to be effected. Chaiphongpachara and Sumruayphol (2017) carried out a study of coastal mosquitoes in an area containing mangroves in Samut Songkhram province, a central province lying southwest of Bangkok, Thailand. The most abundant species were *Anopheles epiroticus* (37.13%), *Culex sitiens* (34.92%), and *C. quinquefasciatus* (27.66%) ($n = 488$) with *C. quinquefasciatus* being the main vector of *Wuchereria bancrofti* and *Brugia malayi* in Southeast Asia (Dickson et al. 2017), *A. epiroticus* is considered to be a secondary vector of malaria, with particular relevance in coastal regions (Sumruayphol et al. 2010; Tananchai et al. 2019). It occurs in coastal brackish water sites from southern Vietnam to peninsular Malaysia (Manguin et al. 2008). The vector capacity for human pathogens by *Culex sitiens* remains insufficiently studied. Other species found in this study were *Aedes aegypti* and *Culex gelidus*. Only *A. epiroticus* and *C. sitiens* were found in significant numbers within 200 m of the sea; however, both species were negatively correlated with mangroves (*A. epiroticus* significantly), *A. epiroticus* positively, significantly correlated with the presence of green algae and *C. sitiens* with temporary but not permanent pools. *C. sitiens* was found to be the dominant brackish water species in other studies in Malaysia (Ismail et al. 2018) and Thailand (Prummongkol et al. 2012). At least for these species, the maintenance of mangrove areas may inhibit the development of vector-mosquito populations.

## 9.5 Tropical Forests

The forests of Southeast Asia are mostly aseasonal. Large-scale seasonal variations in both temperature and rainfall influence tree phenology and species distributions in the marginal tropical forests, whereas seasonality in rainfall is the influential factor in the monsoon tropics (Deb et al. 2018). The impact of climate change is already being felt with increasing tree mortality leading to changes in forest structure and species composition (Hérault and Gourlet-Fleury 2016; Margrove et al. 2015). Deb et al. (2018) indicate that the forest ecosystems of Southeast Asia are highly vulnerable to climate change.

Tropical Southeast Asian forests are amongst the world's most biodiverse areas (Olson and Dinerstein 2002). They are currently undergoing an era of unprecedented destruction (Laurance 2007; Kim et al. 2015; Chisholm et al. 2016; Namkhan et al. 2021) with a subsequent loss of biodiversity (Sodhi et al. 2004; Estoque et al. 2019).

This will be enhanced by climate change with drought increasing in some areas and with the increasing likelihood of extreme weather events and wildfires; the destruction of these forests will also feed back to increase global warming (Chen et al. 2017; Mitchard 2018).

Much of the work on wildfires in Southeast Asia deals with those occurring in peat deposits (Page and Hooijer 2016; Lin et al. 2019); however, forest ecosystems are also in danger (Brando et al. 2019).

Changes in forest habitat will influence a variety of parasites, for example changes in the tick species present are likely to occur with forest destruction with concurrent changes in host availability and microclimate, as well as with an increasing tendency towards drought conditions (Saijuntha et al. 2021a). When forest is converted to agricultural land, particularly for grazing, the tick community will change towards those species using stock animals as hosts and away from those using native animals. Flea and louse communities are likely to be influenced in a similar way due to a reduction in available host species.

Disease vectors also respond to changes in forest systems: one of the main vectors of malaria in Thailand, *Anopheles dirus* has a preferred larval habitat of shallow, slow flowing streams in forest habitats with adaptations to temporary habitats (Rampa Rattanaritikul et al. 1995; Obsomer et al. 2007). It is the major vector of forest malaria in many areas of mainland Southeast Asia (Oo et al. 2003; Sanh et al. 2008; Maeno 2017; Nguitragool et al. 2019). With the destruction of large forested areas, and increasing temperatures and the likelihood of drought driven by climate change the natural habitat of this species is likely to be reduced. With the removal of forest, particularly for agriculture, other vectors of malaria are likely to dominate (Rampa Rattanaritikul et al. 1995).

There is evidence that primates from forests damaged by extreme climate events show a higher level of parasitism those in undisturbed forests (Behie et al. 2014).

## 9.6 Cities, Towns, and Villages

These accumulations of humans are the most likely areas in which direct parasite transmission can occur as the contact rate between individuals and the likelihood of hand to mouth transmission (particularly where sanitary hygiene is poor) are high. In addition, urban areas provide the natural habitat for a variety of human associated hosts such as rats (Zain et al. 2012). Clearing vegetation from such areas will also increase incident radiation and asphalted roads and reflection from the facades of buildings will contribute to the development of heat islands with temperature increases caused by both global warming and land use change (Lee et al. 2017). In addition there is likely to be a scarcity of clean water and an accumulation of gray (waste) water (McIntosh 2014).

Increased temperatures combined with urban drought and the predominance of wastewater will likely change the composition of parasite vector communities (Petney and Taraschewski 2011). Wastewater with a high content of organic matter provides a suitable habitat for *C. quinquefasciatus*, a major vector of *W. bancrofti* (Simonsen and Mwakitalu 2013) and can also act as suitable habitat for various helminth species (Ngoen-klan et al. 2010).

*Ascaris lumbricoides* probably has sufficient resistance to temperature and dry environments to continue to successfully inhabit Southeast Asian cities (Blum and Hotez 2018). Other helminths of public health significance can infect rodent species. Zain et al. (2012) found a community of 11 helminth species (seven nematodes, three cestodes and one acanthocephalan) from 450 *Rattus rattus* and *R. norvegicus* in Kuala Lumpur, Malaysia. Of these *Hymenolepis nana* is a well-known human pathogen (Sirivichayakul et al. 2000; Thompson 2015) and *Angiostrongylus malayensis* is possibly a human pathogen (Prociv et al. 2000).

Hookworms are common parasites, particularly in the tropics and subtropics. *Necator americanus* and *Ancylostoma duodenale* occur worldwide, while *A. ceylonensis* is also found in Southeast Asia with *N. americanus* and *A. ceylonensis* occurring most frequently depending on location (Jiraanankul et al. 2011; Ngui et al. 2012; Inpankaew et al. 2014). Climate change is thought to be a possible cause for the switching of species dominance. For example, the dominance of *N. americanus* may switch to an *Ancylostoma* species due to increasing temperatures and drought as *Ancylostoma* species can undergo arrested developmental as perpetual larvae in human tissues, thus surviving environmental extremes (Blum and Hotez 2018).

## 9.7 Agricultural Land

There have been major increases in agricultural land throughout Southeast Asia (Imai et al. 2018; Zeng et al. 2018) with considerable implications for human health (Burkett-Cadena and Vittor 2018; Shah et al. 2019). Much of the change occurs via forest destruction, dominated by increases in wood and food production. The destruction of forests opens land to increased incident radiation and thus to both flooding and drought potential (Prăvălie 2018), as well as releasing carbon into the atmosphere feeding back positively on global warming (Lawrence and Vandecar 2015). The area under a variety of high yield cash crops, such as rubber, oil palm, and coffee, has increased massively over the last two decades (Fox and Castella 2013).

Shah et al. (2019) performed a meta-analysis which showed that people exposed to an agricultural setting in Southeast Asia are on average 1.74 (CI 1.47–2.07) times more likely to be infected with a pathogen than those unexposed, with oil palm, rubber, and non-poultry based livestock farming being significantly associated with hookworm and malaria infections.

Rubber plantations have shown unprecedented growth (Ziegler et al. 2009). In 2009, Petney et al. predicted that such plantations would offer the shade and small streams ideal as breeding sites for *Anopheles dirus*, a major vector of malaria in Thailand. Since this time a variety of malaria vectors have been collected in rubber plantations (*An. aconitus*, *An. barbirostris* s.l., *An. campestris*, *An. dirus* s.l., An. latens, *An. maculatus* s.l., *An. minimus* s.l., *An. philippinensis*, *An. pseudowillmori*, *Anopheles umbrosus* sensu lato (s.l.)), with their significance depending on the location of the plantation, its age and the season of collection (Tangena et al. 2016; Sumarnrote et al. 2017; Pimnon and Bhumiratana 2018).

Thellmann et al. (2019) have shown that climate change leading to particularly dry years could have a significant effect on the water balance via higher evapotranspiration rates increasing periods of water scarcity. Nevertheless, for the Greater Mekong Subregion the area suitable for rubber cultivation will potentially increase to 50% in 2030 compared to 44.3% at the turn of the century (Golbon et al. 2018).

The transmission of filariasis is also potentially influenced by climate change. Of the filarial species in Southeast Asia *Brugia malayi* is the predominant species Brunei, Darussalam, Indonesia, Malaysia, and Vietnam while *W. bancrofti* predominates in Lao PDR, Philippines, and Myanmar. Both species occur in Cambodia, the southern Philippine islands, and Thailand (Noordin et al. 2013). A third species, *B. timori*, is confined to the Lesser Sunda Islands on Indonesia (Fischer et al. 2004).

In the Philippines *Mansonia uniformis* and *Ma. bonneae*, both vectors of *B. malayi*, breed in rice paddies (Cabrera and Jueco 1972). Rice fields are also a significant habitat for the life cycle of trematodes with snails as intermediate hosts. The liver fluke *Opisthorchis viverrini* (Kiatsopit et al. 2012) is the major risk factor for developing cholangiocarcinoma in the Lower Mekong Region (Khuntikeo et al. 2018a). This is a major cause of mortality and socio-economic distress for rural families in this area (Khuntikeo et al. 2018b). The success of the life cycle of the parasite in rice fields is enhanced by rice farmers buying fish for rice-fish agriculture, with the likelihood that the fish supplied by hatcheries and nurseries and which act as second intermediate hosts are already infected (Pitaksakulrat et al. 2013; Oron et al. 2015). The *Bithynia* snail first intermediate hosts already occur in most freshwater ecosystems throughout the area (Petney et al. 2013).

In addition, other trematode pathogens can be found in snail intermediate hosts in rice fields, including *Haplorchis taichui*, *H. pumilio* and various echinostome species (Saijuntha et al. 2013; Madsen et al. 2015; Sato et al. 2015).

How will climate change influence the diversity of rice field parasites? Three factors come into play, increasing temperature, reduced or increased rainfall and extreme weather events. Estimates suggest that rice yield will continue to be reduced as climate change increases (Prabnakorn et al. 2018, 2019). Increasing temperature will most likely speed up the life cycle of the ectothermic intermediate hosts, as well as the parasites, potentially leading to more rapid population growth. Decreasing rainfall and increasing drought conditions, for example in northeast Thailand, may reduce suitable habitat for the intermediate hosts thus reducing population levels. This is particularly true for species with limited salt tolerance, but may be an advantage for species preferring more saline conditions (Kim et al. 2016; Hintz and Relyea 2019) providing salinity does not rise above lethal thresholds as the northeast of Thailand lies on a saline geological formation and water salinity will increase with climate change (Pholkern et al. 2018). The survival of *O. viverrini* eggs is also dependent on salinity, above a threshold of which the eggs do not survive (Sereewong et al. 2018).

The effect of drought was determined for *Bithynia siamensis goniomphalos*, the first intermediate host of *Opisthorchis viverrini*. This led to a population collapse with the drying out of its pond habitat, with few specimens surviving deep in the remaining mud. Infection rates with *O. viverrini* plummeted during this time. With sufficient rainfall, the population of *B. s. goniomphalos* increased rapidly (Brockelman et al. 1986; Chaiyasaeng et al. 2019).

Drought is also likely to increase soil salinization, particularly in areas such as northeast Thailand where sub-surface salt layers occur (Kohyama et al. 1993). This can influence land-use and subsequently parasite host distribution.

In Southeast Asia three members of the genus *Schistosoma* that are pathogens of humans are known, *S. japonicum*, *S. malayensis*, and *S. mekongi*. *Schistosoma japonicum* is the most common pathogen of humans in parts of Southeast Asia and China, and is the only human pathogenic schistosome in the Philippines and the Celebes (Indonesia) (Yang and Bergquist 2018). It is frequently found together with its snail intermediate hosts in rice fields (Abao-Paylangco et al. 2019). Models based on climate change and the biology of *S. japonicum* show an increased risk of infection throughout the region (Yang and Bergquist 2018). *Schistosoma malayensis* is only found in parts of the Malay Peninsula, while *S. mekongi* occurs in parts of Cambodia, Lao PDR, and Thailand (Gordon et al. 2019).

## 9.8 Natural Freshwater Ecosystems

Southeast Asia has some of the most extensive, endangered, wetland ecosystems in the world (Dudgeon 2002) containing very high levels of biodiversity (Abell et al. 2008). These wetlands are currently under threat from a number of factors including human population growth and the conversion of natural habitats to agricultural land (destruction, extensive water use, pollution, canalization, dam construction), as well as transborder conflicts, all of which will be influenced by climate change, particularly drought and extreme weather events (Dudgeon 2002, 2019; Welcomme et al. 2016; Williams and Patricola 2018). These changes are likely to have a significant effect on the transmission of parasites with freshwater aspects in their life cycles (Sithithaworn et al. 2012; Pitaksakulrat et al. 2013; Ziegler et al. 2013). As indicated above, the increasing human population is not only reducing the extent of natural freshwater ecosystems via land-use change, but also increasing the amount of wastewater, potentially contaminated with parasites, into such systems (Zhao et al. 2006; Evans et al. 2012).

Freshwater fish and crustaceans are a major protein source in inland areas of mainland Southeast Asia where it is traditional to eat them either raw, fermented, and generally undercooked (Petney et al. 2018; Saijuntha et al. 2019). The breeding season of many commercially important species is coordinated with the rise in water level in the monsoon season during which time they move from the rivers into the

**Fig. 9.2** Preliminary construction of fish traps (Lob yuen) along the Songkhram River in northeast Thailand. The water flow is from left to right and the water level will rise to above the top horizontal bar on the construction. The fish move into the dense bamboo vegetation for protection during the breeding season. (Photo T.N. Petney)

flood plains. As the water recedes they move back into rivers and streams (Tanaka et al. 2015). These organisms are intermediate hosts of a variety of parasite species pathogenic to humans, for example *Opisthorchis viverrini* a liver fluke that is the major risk factor for human cholangiocarcinoma in the region (Khuntikeo et al. 2018a), *Haplorchis* (Onsurathum et al. 2016), and *Paragonimus* species (Yoshida et al. 2019). Figure 9.2 shows bamboo constructions well above the water level of the Songkhram River to which fish traps are tied during the rainy season.

Communities of parasites and hosts can be extensive, for example a large number of freshwater snails act as hosts for a wide variety of trematodes of both medical and veterinary importance (Saijuntha et al. 2021b). Mard-arhin et al. (2001) examined the parasite community in snails and fish from natural freshwater ecosystems in northern Thailand. They found eight taxa of helminth parasite none of which could be identified to species level. A variety of other studies have found similar results with relatively high parasite diversity in such natural systems (Saenphet et al. 2001; Wongsawad et al. 2004; Kiatsopit et al. 2016; Veeravechsukij et al. 2018).

The north and northeast of Thailand are subject to droughts that are likely to increase with climate change, potentially leading to increases in salinity in both urban and rural areas (Arjwech et al. 2019). The salt originates from lacustrine salt

**Fig. 9.3** Salt deposited on the soil surface in Nakhon Ratchasima province in Northeast-Thailand. The photograph is courtesy of the German Federal Institute for Geosciences and Natural Resources from the project "Salinisation of soils in the province of Nakhon Ratchasima (Northeast-Thailand)"

layers lying between layers of shale and sandstone (Tabakh et al. 1998). These salt-affected soils limit drainage during the monsoon season, during which the salt moves towards the soil surface where it deposits via evaporation during the dry season (Fig. 9.3) (Sinanuwong and Takaya 1974). High surface salt concentrations leaching into freshwater systems during the monsoon season will likely to influence the distribution and longevity of intermediate hosts as well as their cercariae, for example *O. viverrini* (see above).

Southeast Asia is one of the regions most likely to be affected by sea level rise worldwide (see above, Schuerch et al. 2018). Sea level rise will convert near coastal freshwater habitats into brackish or saltwater habitats (Ahmed et al. 2019), reducing the risk of infection with parasites associated with freshwater, but increasing the risk, for example of malaria, transmitted by salt or brackish water vectors (Sumruayphol et al. 2010; Tananchai et al. 2019). As with agricultural rice field systems, a variety of natural wetland systems also act as habitats for parasites. In the Philippines *S. japonicum* and its snail intermediate hosts are also found on wet soil surfaces, swamps, ponds, and streams (Gordon et al. 2019).

## 9.9 Evolution and Adaptation

Superimposed on all of these factors is the possibility of short-term evolutionary changes leading to the adaptation of species to the new environmental conditions imposed on them (Hoffmann and Sgro 2011; Bush et al. 2016; Nogués-Bravo et al. 2018; Kelly 2019) or shifts due to phenotypic plasticity (Hoberg and Brooks 2015). Evolution can occur over short periods of time in the face of environmental change (ecological evolution) (Thompson 1998; Schoener 2011), a phenomenon also found in parasites (Kelehear et al. 2012; Weclawski et al. 2013, 2014) and hosts (Duffy et al. 2009; Mateos-Gonzalez et al. 2015). The significance of host-parasite interactions in relation to climate change remains to be studied in the field in detail for a sufficient time (Sibley 2019).

## 9.10 Conclusions

Climate change will have a significant impact on parasites and parasite communities that respond to increasing temperatures, extreme weather events and changes in the hydrodynamics of the area in which they live. Southeast Asia represents a spatial mosaic with some areas likely to experience increasing drought conditions while other will experience increased precipitation. However, most of the predictions both in relation to local climate and to the parasites' response to climate change are based on models. What are urgently needed are long-term case studies tracking climate in association with parasite diversity as well as the dynamics and distribution of individual parasite and host species. In addition, comparatively little is known about "non-pathogenic" parasites of wild animals, indicating that our knowledge of the true biodiversity of parasites is inadequate.

**Acknowledgements** Dr. Tanapipat Walalite kindly supplied information relevant to Fig. 9.2. The German Federal Institute for Geosciences and Natural Resources kindly allowed publication of Fig. 9.3.

## References

Abao-Paylangco RA, Balamad MK, Paylangco JC, Japitana RA, Jumawan JC (2019) Schistosoma japonicum in selected rice fields surrounding Lake Mainit, Philippines. J Ecosyst Sci Eco-Gov 1:15–24

Abell R, Thieme ML, Revenga C, Bryer M, Kottelat M, Bogutskaya N, Coad B, Mandrak N, Balderas SC, Bussing W, Stiassny ML (2008) Freshwater ecoregions of the world: a new map of biogeographic units for freshwater biodiversity conservation. Bioscience 58(5):403–414

Ahmed N, Thompson S, Glaser M (2019) Global aquaculture productivity, environmental sustainability, and climate change adaptability. Environ Manag 63(2):159–172

Altizer S, Ostfeld RS, Johnson PT, Kutz S, Harvell CD (2013) Climate change and infectious diseases: from evidence to a predictive framework. Science 341(6145):514–519

Amnuaylojaroen T, Chanvichit P (2019) Projection of near-future climate change and agricultural drought in Mainland Southeast Asia under RCP8.5. Clim Chang 155:175–193

Anh VT, Tram NT, Klank LT, Cam PD, Dalsgaard A (2007) Faecal and protozoan parasite contamination of water spinach (*Ipomoea aquatica*) cultivated in urban wastewater in Phnom Penh, Cambodia. Tropical Med Int Health 12:73–81

Anticamara JA, Go KT (2017) Impacts of super-typhoon Yolanda on Philippine reefs and communities. Reg Environ Change 17(3):703–713

Arjwech R, Everett ME, Schulmeister MK (2019) Soil salinity mapping of an urbanizing area in Northeast Thailand. Q J Eng Geol Hydrogeol 17:qjegh2019-113

Bamber JL, Oppenheimer M, Kopp RE, Aspinall WP, Cooke RM (2019) Ice sheet contributions to future sea-level rise from structured expert judgment. Proc Natl Acad Sci 116 (23):11195–111200

Behie AM, Kutz S, Pavelka MS (2014) Cascading effects of climate change: do hurricane-damaged forests increase risk of exposure to parasites? Biotropica 46(1):25–31

Blum AJ, Hotez PJ (2018) Global "worming": climate change and its projected general impact on human helminth infections. PLoS Negl Trop Dis 12(7):e0006370. https://doi.org/10.1371/journal.pntd.0006370

Brando PM, Paolucci L, Ummenhofer CC, Ordway EM, Hartmann H, Cattau ME, Rattis L, Medjibe V, Coe MT, Balch J (2019) Droughts, wildfires, and forest carbon cycling: A pantropical synthesis. Annu Rev Earth Planet Sci 47:555–581

Brockelman WY, Upatham ES, Viyanant V, Ardsungnoen S, Chantanawat R (1986) Field studies on the transmission of the human liver fluke, *Opisthorchis viverrini*, in northeast Thailand: population changes of the snail intermediate host. Int J Parasitol 16:545–552

Brooks DR, Hoberg EP (2007) How will global climate change affect parasite–host assemblages? Trends Parasitol 23(12):571–574

Burkett-Cadena ND, Vittor AY (2018) Deforestation and vector-borne disease: forest conversion favors important mosquito vectors of human pathogens. Basic Appl Ecol 26:101–110

Bush A, Mokany K, Catullo R, Hoffmann A, Kellermann V, Sgrò C, McEvey S, Ferrier S (2016) Incorporating evolutionary adaptation in species distribution modelling reduces projected vulnerability to climate change. Ecol Lett 19(12):1468–1478

Cabrera BD, Jueco NL (1972) Filariasis survey among indigenous tribes of Palawan, Republic of the Philippines. Southeast Asian J Trop Med Public Health 3(1):31–39

Cabrera JS, Lee HS (2018) Impacts of climate change on flood-prone areas in Davao Oriental, Philippines. Water 10(7):893

Carlson CJ, Burgio KR, Dougherty ER, Phillips AJ, Bueno VM, Clements CF, Castaldo G, Dallas TA, Cizauskas CA, Cumming GS, Doña J (2017) Parasite biodiversity faces extinction and redistribution in a changing climate. Sci Adv 3(9):e1602422

Chaiphongpachara T, Sumruayphol S (2017) Species diversity and distribution of mosquito vectors in coastal habitats of Samut Songkhram province, Thailand. Trop Biomed 34(3):524–532

Chaiyasaeng M, Pechdee P, Sereewong C, Suwannatrai A, Laha T, Tesana S (2019) Effects of aestivation on survival of *Bithynia siamensis goniomphalos* snails and the infection of *Opisthorchis viverrini* in the irrigation area of wet-and dry-season rice paddy. Acta Trop 192:55–60

Chandrasa GT, Montenegro A (2020) Evaluation of regional climate model simulated rainfall over Indonesia and its application for downscaling future climate projections. Int J Climatol 40 (4):2026–2047

Chen Y, Morton DC, Andela N, Van Der Werf GR, Giglio L, Randerson JT (2017) A pan-tropical cascade of fire driven by El Niño/southern oscillation. Nat Clim Chang 7(12):906–911

Chevalier V, Courtin F, Guis H, Tran A, Vial L (2016) Climate change and vector-borne diseases. In: Torquebiau E (ed) Climate change and agriculture worldwide. Springer, Dordrecht, pp 97–108

Chisholm RA, Wijedasa LS, Swinfield T (2016) The need for long-term remedies for Indonesia's forest fires. Cons Biol 30(1):5–6

Church JA, Clark PU, Cazenave A, Gregory JM, Jevrejeva S, Levermann A, Merrifield MA, Milne GA, Nerem RS, Nunn PD, Payne AJ (2013) Sea level change. In: Climate change 2013: the physical science basis. Contribution of working group I to the fifth assessment report of the intergovernmental panel on climate change. IPCC, Geneva, p 1137

Cizauskas CA, Carlson CJ, Burgio KR, Clements CF, Dougherty ER, Harris NC, Phillips AJ (2017) Parasite vulnerability to climate change: an evidence-based functional trait approach. R Soc Open Sci 4(1):160535

Clausen JH, Madsen H, Van PT, Dalsgaard A, Murrell KD (2015) Integrated parasite management: path to sustainable control of fishborne trematodes in aquaculture. Trends Parasitol 31(1):8–15

Deb JC, Phinn S, Butt N, McAlpine CA (2018) Climate change impacts on tropical forests: identifying risks for tropical Asia. J Trop Forest Sci 30(2):182–194

Dickson BF, Graves PM, McBride WJ (2017) Lymphatic filariasis in mainland Southeast Asia: a systematic review and meta-analysis of prevalence and disease burden. Trop Med Inf Dis 2 (3):32

dos Santos CA, Howgate P (2011) Fishborne zoonotic parasites and aquaculture: a review. Aquaculture 318(3–4):253–261

Dudgeon D (2002) The most endangered ecosystems in the world? Conservation of riverine biodiversity in Asia. Int Verein Theoret Angew Limnol Verhandl 28(1):59–68

Dudgeon D (2019) Multiple threats imperil freshwater biodiversity in the Anthropocene. Curr Biol 29(19):R960–R967

Duffy MA, Hall SR, Cáceres CE, Ives AR (2009) Rapid evolution, seasonality, and the termination of parasite epidemics. Ecology 90(6):1441–1448

Erban LE, Gorelick SM, Zebker HA (2014) Groundwater extraction, land subsidence, and sea-level rise in the Mekong Delta, Vietnam. Environ Res Lett 9(8):084010

Estoque RC, Ooba M, Avitabile V, Hijioka Y, DasGupta R, Togawa T, Murayama Y (2019) The future of Southeast Asia's forests. Nat Commun 10(1):1–2

Evans AE, Hanjra MA, Jiang Y, Qadir M, Drechsel P (2012) Water quality: assessment of the current situation in Asia. Int J Water Res Dev 28(2):195–216

Fischer P, Supali T, Maizels RM (2004) Lymphatic filariasis and *Brugia timori*: prospects for elimination. Trends Parasitol 20(8):351–355

Forster PM, Maycock AC, McKenna CM, Smith CJ (2020) Latest climate models confirm need for urgent mitigation. Nat Clim Chang 10(1):7–10

Fox J, Castella JC (2013) Expansion of rubber (*Hevea brasiliensis*) in Mainland Southeast Asia: what are the prospects for smallholders? J Peasant Stud 40(1):155–170

Freund MB, Henley BJ, Karoly DJ, McGregor HV, Abram NJ, Dommenget D (2019) Higher frequency of Central Pacific El Niño events in recent decades relative to past centuries. Nat Geosci 12(6):450–455

Ge F, Zhu S, Peng T, Zhao Y, Sielmann F, Fraedrich K, Zhi X, Liu X, Tang W, Ji L (2019) Risks of precipitation extremes over Southeast Asia: does 1.5° C or 2° C global warming make a difference? Env. Res Lett 14(4):044015

Golbon R, Cotter M, Sauerborn J (2018) Climate change impact assessment on the potential rubber cultivating area in the Greater Mekong Subregion. Environ Res Lett 13(8):084002

Gordon CA, Kurscheid J, Williams GM, Clements AC, Li Y, Zhou XN, Utzinger J, McManus DP, Gray DJ (2019) Asian schistosomiasis: current status and prospects for control leading to elimination. Trop Med Inf Dis 4(1):40

Hérault B, Gourlet-Fleury S (2016) Will tropical rainforests survive climate change? In: Torquebiau E (ed) Climate change and agriculture worldwide. Springer, Dordrecht, pp 183–196

Himsworth CG, Parsons KL, Jardine C, Patrick DM (2013) Rats, cities, people, and pathogens: a systematic review and narrative synthesis of literature regarding the ecology of rat-associated zoonoses in urban centers. Vector-Borne Zoont Dis 13(6):349–359

Hintz WD, Relyea RA (2019) A review of the species, community, and ecosystem impacts of road salt salinisation in fresh waters. Freshw Biol 64(6):1081–1097

Hoberg EP, Brooks DR (2015) Evolution in action: climate change, biodiversity dynamics and emerging infectious disease. Phil Trans R Soc B Biol Sci 370(1665):20130553

Hoffmann AA, Sgro CM (2011) Climate change and evolutionary adaptation. Nature 470 (7335):479–485

Imai N, Furukawa T, Tsujino R, Kitamura S, Yumoto T (2018) Factors affecting forest area change in Southeast Asia during 1980-2010. PLoS One 13(5):e0197391

Inpankaew T, Schär F, Dalsgaard A, Khieu V, Chimnoi W, Chhoun C, Sok D, Marti H, Muth S, Odermatt P, Traub RJ (2014) High prevalence of *Ancylostoma ceylanicum* hookworm infections in humans, Cambodia, 2012. Emerg Inf Dis 20(6):976

IPCC (2019) Summary for policymakers. In: Shukla PR, Skea J, Calvo Buendia E, Masson-Delmotte V, Pörtner H-O, Roberts DC, Zhai P, Slade R, Connors S, van Diemen R, Ferrat M, Haughey E, Luz S, Neogi S, Pathak M, Petzold J, Portugal Pereira J, Vyas P, Huntley E, Kissick K, Belkacemi M, Malley J (eds) Climate change and land: an IPCC special report on climate change, desertification, land degradation, sustainable land management, food security, and greenhouse gas fluxes in terrestrial ecosystems. IPCC, Geneva

Ismail NW, Chan SM (2019) Impacts of the El Niño-southern oscillation (ENSO) on paddy production in Southeast Asia. Clim Dev 5:1–3

Ismail TN, A Kassim NF, A Rahman A, Yahya K, Webb CE (2018) Day biting habits of mosquitoes associated with mangrove forests in Kedah, Malaysia. Trop Med Inf Dis 3(3):77

Jevrejeva S, Jackson LP, Riva RE, Grinsted A, Moore JC (2016) Coastal sea level rise with warming above 2 C. Proc Natl Acad Sci 113(47):13342–13347

Jia G, Shevliakova E, Artaxo P, De Noblet-Ducoudré N, Houghton R, House J, Kitajima K, Lennard C, Popp A, Sirin A, Sukumar R, Verchot L (2019) Land-climate interactions. In: Shukla PR, Skea J, Buendia EC, Masson-Delmotte V, Pörtner H-O, Roberts DC, Zhai P, Slade R, Connors S, van Diemen R, Ferrat M, Haughey E, Luz S, Neogi S, Pathak M, Petzold J, Pereira JP, Vyas P, Huntley E, Kissick K, Belkacemi M, Malley J (eds) Climate change and land: an IPCC special report on climate change, desertification, land degradation, sustainable land management, food security, and greenhouse gas fluxes in terrestrial ecosystems. IPCC, Geneva

Jiraanankul V, Aphijirawat W, Mungthin M, Khositnithikul R, Rangsin R, Traub RJ, Piyaraj P, Naaglor T, Taamasri P, Leelayoova S (2011) Incidence and risk factors of hookworm infection in a rural community of central Thailand. Am J Trop Med Hyg 84(4):594–598

Kang S, Im ES, Eltahir EA (2019) Future climate change enhances rainfall seasonality in a regional model of western maritime continent. Clim Dyn 52(1–2):747–764

Kelehear C, Brown GP, Shine R (2012) Rapid evolution of parasite life history traits on an expanding range-edge. Ecol Lett 15(4):329–337

Kelly M (2019) Adaptation to climate change through genetic accommodation and assimilation of plastic phenotypes. Phil Trans R Soc B 374(1768):20180176

Khuntikeo N, Thinkhamrop B, Bundhamcharoen K, Andrews RH, Grundy-Warr C, Yongvanit P, Loilome W, Chamadol N, Kosuwan W, Sithithaworn P, Petney TN (2018a) The socioeconomic burden of cholangiocarcinoma associated with *Opisthorchis viverrini* sensu lato infection in Northeast Thailand: a preliminary analysis. Adv Parasitol 102:141–163

Khuntikeo N, Titapun A, Loilome W, Yongvanit P, Thinkhamrop B, Chamadol N, Boonmars T, Nethanomsak T, Andrews RH, Petney TN, Sithithaworn P (2018b) Current perspectives on opisthorchiasis control and cholangiocarcinoma detection in Southeast Asia. Front Med 5:117

Kiatsopit N, Sithithaworn P, Saijuntha W, Boonmars T, Tesana S, Sithithaworn J, Petney TN, Andrews RH (2012) Exceptionally high prevalence of infection of *Bithynia siamensis goniomphalos* with *Opisthorchis viverrini* cercariae in different wetlands in Thailand and Lao PDR. Am J Trop Med Hyg 86(3):464–469

Kiatsopit N, Sithithaworn P, Kopolrat K, Namsanor J, Andrews RH, Petney TN (2016) Trematode diversity in the freshwater snail *Bithynia siamensis goniomphalos* sensu lato from Thailand and Lao PDR. J Helminthol 90(3):312–320

Kim DH, Sexton JO, Townshend JR (2015) Accelerated deforestation in the humid tropics from the 1990s to the 2000s. Geophys Res Lett 42(9):3495–3501

Kim CS, Echaubard P, Suwannatrai A, Kaewkes S, Wilcox BA, Sripa B (2016) Seasonal and spatial environmental influence on *Opisthorchis viverrini* intermediate hosts, abundance, and distribution: insights on transmission dynamics and sustainable control. PLoS Negl Trop Dis 10(11): e0005121

King AD, Karoly DJ, van Oldenborgh GJ (2016) Climate change and El Niño increase likelihood of Indonesian heat and drought. Bull Am Meteorol Soc 97(12):S113–S117

Kohyama K, Wichaidit P, Pramojanee P, Sukchan S, Hidenori W (1993) Salinization in the watershed of Northeast Thailand. In: Kohyama K, Subhasaram T (eds) Salt-affected soils in Northeast Thailand: their salinization and amelioration. Agricultural Development Research Center, Japanese International Cooperation Agency, Khon Kaen, pp 1–26

Laurance WF (2007) Forest destruction in tropical Asia. Curr Sci 10:1544–1550

Lawrence D, Vandecar K (2015) Effects of tropical deforestation on climate and agriculture. Nat Clim 5(1):27–36

Lee HS, Trihamdani AR, Kubota T, Iizuka S, Phuong TT (2017) Impacts of land use changes from the Hanoi master plan 2030 on urban heat islands: part 2. Influence of global warming. Sustain Cities Soc 31:95–108

Li XX (2020) Heat wave trends in Southeast Asia during 1979–2018: the impact of humidity. Sci Total 5:137664

Lin S, Sun P, Huang X (2019) Can peat soil support a flaming wildfire? Int J Wildland Fire 28 (8):601–613

Lin TC, Hogan JA, Chang CT (2020) Tropical cyclone ecology: A scale-link perspective. Trends Ecol Evol 35:594–604

Madsen H, Dung BT, Viet NK, Dalsgaard A, Van PT (2015) The role of rice fields, fish ponds and water canals for transmission of fish-borne zoonotic trematodes in aquaculture ponds in Nam Dinh Province, Vietnam. Parasites Vector 8(1):625

Maeno Y (2017) Molecular epidemiology of mosquitoes for the transmission of forest malaria in south-central Vietnam. Trop Med Health 45(1):27

Manguin S, Garros C, Dusfour I, Harbach RE, Coosemans M (2008) Bionomics, taxonomy, and distribution of the major malaria vector taxa of *Anopheles* subgenus *Cellia* in Southeast Asia: an updated review. Inf Genet Evol 8(4):489–503

Marcogliese DJ (2008) The impact of climate change on the parasites and infectious diseases of aquatic animals. Rev Sci Tech 27(2):467–484

Marcogliese DJ (2016) The distribution and abundance of parasites in aquatic ecosystems in a changing climate: more than just temperature. Integr Comp Biol 56(4):611–619

Mard-Arhin N, Prawang T, Wongsawad C (2001) Helminths of freshwater animals from five provinces in northern Thailand. Southeast Asian J Trop Med Public Health 32:206–209

Margrove JA, Burslem DF, Ghazoul J, Khoo E, Kettle CJ, Maycock CR (2015) Impacts of an extreme precipitation event on dipterocarp mortality and habitat filtering in a Bornean tropical rain forest. Biotropica 47(1):66–76

Mas-Coma S, Valero MA, Bargues MD (2009) Climate change effects on trematodiases, with emphasis on zoonotic fascioliasis and schistosomiasis. Vet Parasitol 163(4):264–280

Mateos-Gonzalez F, Sundström LF, Schmid M, Björklund M (2015) Rapid evolution of parasite resistance in a warmer environment: insights from a large scale field experiment. PLoS One 10 (6):e0128860

McIntosh AC (2014) Urban water supply and sanitation in Southeast Asia: a guide to good practice. Asian Development Bank, Manila

Mitchard ET (2018) The tropical forest carbon cycle and climate change. Nature 559 (7715):527–354

Molnár PK, Dobson AP, Kutz SJ (2013) Gimme shelter–the relative sensitivity of parasitic nematodes with direct and indirect life cycles to climate change. Glob Chang Biol 19 (11):3291–3305

Namkhan M, Gale GA, Savini T, Tantipisanuh N (2021) Loss and vulnerability of lowland forests in mainland Southeast Asia. Conserv Biol 35:206–215

Ngai ST, Juneng L, Tangang F, Chung JX, Salimun E, Tan ML, Amalia S (2020) Future projections of Malaysia daily precipitation characteristics using bias correction technique. Atmos Res 240:104926

Ngoen-klan R, Piangjai S, Somwang P, Moophayak K, Sukontason K, Sukontason KL, Sampson M, Irvine K (2010) Emerging helminths infection in snails and cyprinoid fish in sewage treatment wetlands waters in Cambodia. Asian J Water Environ Pollut 7(3):13–21

Ngui R, Ching LS, Kai TT, Roslan MA, Lim YA (2012) Molecular identification of human hookworm infections in economically disadvantaged communities in peninsular Malaysia. Am J Trop Med Hyg 86(5):837–842

Nguitragool W, Karl S, White M, Koepfli C, Felger I, Singhasivanon P, Mueller I, Sattabongkot J (2019) Highly heterogeneous residual malaria risk in western Thailand. Int J Parasitol 49 (6):455–462

Nogués-Bravo D, Rodríguez-Sánchez F, Orsini L, De Boer E, Jansson R, Morlon H, Fordham DA, Jackson ST (2018) Cracking the code of biodiversity responses to past climate change. Trends Ecol Evol 33(10):765–776

Noordin R, Shenoy RK, Lim BH, Ramachandran CP (2013) Filarial worms in Southeast Asia. In: Lim YA, Vythilingam I (eds) Parasites and their vectors: a special focus on Southeast Asia. Springer, Vienna, pp 33–56

Nunez S, Arets E, Alkemade R, Verwer C, Leemans R (2019) Assessing the impacts of climate change on biodiversity: is below 2° C enough? Clim Chang 154(3–4):351–365

Obsomer V, Defourny P, Coosemans M (2007) The *Anopheles dirus* complex: spatial distribution and environmental drivers. Malar J 6(1):1–6

Olson DM, Dinerstein E (2002) The global 200: a representation approach to conserving the Earth's most biologically valuable ecoregions. Conserv Biol 12(3):502–515

O'Neill BC, Oppenheimer M, Warren R, Hallegatte S, Kopp RE, Pörtner HO, Scholes R, Birkmann J, Foden W, Licker R, Mach KJ (2017) IPCC reasons for concern regarding climate change risks. Nat Clim Chang 7(1):28–37

Onsurathum S, Pinlaor P, Charoensuk L, Haonon O, Chaidee A, Intuyod K, Laummaunwai P, Boonmars T, Kaewkes W, Pinlaor S (2016) Contamination of *Opisthorchis viverrini* and *Haplorchis taichui* metacercariae in fermented fish products in northeastern Thailand markets. Food Control 59:493–498

Oo TT, Storch V, Becker N (2003) *Anopheles dirus* and its role in malaria transmission in Myanmar. J Vector Ecol 28:175–183

Oron G, Adel M, Agmon V, Friedler E, Halperin R, Leshem E, Madsen H, Dung BT, Viet NK, Dalsgaard A, Van PT (2015) The role of rice fields, fish ponds and water canals for transmission of fish-borne zoonotic trematodes in aquaculture ponds in Nam Dinh Province, Vietnam. Parasites Vectors 8(1):625

Page SE, Hooijer A (2016) In the line of fire: the peatlands of Southeast Asia. Phil Trans R Soc B Biol Sci 371(1696):20150176

Paull SH, Johnson PT (2011) High temperature enhances host pathology in a snail-trematode system: possible consequences of climate change for the emergence of disease. Freshw Biol 56(4):767–778

Petney TN, Taraschewski H (2011) Water-borne parasitic diseases: hydrology, regional development and control. In: Frimmel FH (ed) Water chemistry and microbiology, vol. 3, A treatise on water science. Elsevier, Amsterdam, pp 303–366

Petney T, Sithithaworn P, Satrawaha R, Grundy-Warr C, Andrews R, Wang YC, Feng CC (2009) Potential malaria reemergence, northeastern Thailand. Emerg Infect Dis 15(8):1330

Petney TN, Andrews RH, Saijuntha W, Wenz-Mücke A, Sithithaworn P (2013) The zoonotic, fish-borne liver flukes *Clonorchis sinensis*, *Opisthorchis felineus* and *Opisthorchis viverrini*. Int J Parasitol 43(12–13):1031–1046

Petney TN, Andrews RH, Saijuntha W, Tesana S, Prasopdee S, Kiatsopit N, Sithithaworn P (2018) Taxonomy, ecology and population genetics of *Opisthorchis viverrini* and its intermediate hosts. Adv Parasitol 101:1–39

Pholkern K, Saraphirom P, Srisuk K (2018) Potential impact of climate change on groundwater resources in the central Huai Luang Basin, Northeast Thailand. Sci Total Environ 633:1518–1535

Pimnon S, Bhumiratana A (2018) Adaptation of *Anopheles* vectors to anthropogenic malaria-associated rubber plantations and indoor residual spraying: establishing population dynamics and insecticide susceptibility. Can J Infect Dis Med Microbiol 2018:9853409

Pitaksakulrat O, Sithithaworn P, Laoprom N, Laha T, Petney TN, Andrews RH (2013) A cross-sectional study on the potential transmission of the carcinogenic liver fluke *Opisthorchis viverrini* and other fishborne zoonotic trematodes by aquaculture fish. Foodborne Pathog Dis 10(1):35–41

Polley L, Thompson RA (2009) Parasite zoonoses and climate change: molecular tools for tracking shifting boundaries. Trends Parasitol 25(6):285–291

Prabnakorn S, Maskey S, Suryadi FX, de Fraiture C (2018) Rice yield in response to climate trends and drought index in the Mun River Basin, Thailand. Sci Total Environ 1621:108–119

Prabnakorn S, Maskey S, Suryadi FX, de Fraiture C (2019) Assessment of drought hazard, exposure, vulnerability, and risk for rice cultivation in the Mun River Basin in Thailand. Nat Hazards 97(2):891–911

Prăvălie R (2018) Major perturbations in the Earth's forest ecosystems. Possible implications for global warming. Earth-Sci Rev 185:544–571

Prociv P, Spratt DM, Carlisle MS (2000) Neuro-angiostrongyliasis: unresolved issues. Int J Parasitol 30(12–13):1295–1303

Prummongkol S, Panasoponkul C, Apiwathnasorn C, Lek-Uthai U (2012) Biology of *Culex sitiens*, a predominant mosquito in Phang Nga, Thailand after a tsunami. J Insect Sci 12:1–8

Qian X, Qiu B, Zhang Y (2019) Widespread decline in vegetation photosynthesis in Southeast Asia due to the prolonged drought during the 2015/2016 El Niño. Remote Sens 11(8):910

Rampa Rattanaritikul I, Green CA, Panyims S (1995) Larval habitats of malaria vectors and other *Anopheles* mosquitoes around a transmission focus in northwestern Thailand. J Am Mosq Control Assoc 1(4):428–433

Rifai SW, Li S, Malhi Y (2019) Coupling of El Niño events and long-term warming leads to pervasive climate extremes in the terrestrial tropics. Environ Res Lett 14(10):105002

Rohr JR, Dobson AP, Johnson PT, Kilpatrick AM, Paull SH, Raffel TR, Ruiz-Moreno D, Thomas MB (2011) Frontiers in climate change–disease research. Trends Ecol Evol 26(6):270–277

Román-Palacios C, Wiens JJ (2020) Recent responses to climate change reveal the drivers of species extinction and survival. Proc Nat Acad Sci 117(8):4211–4217

Sa'adi Z, Shiru MS, Shahid S, Ismail T (2020) Selection of general circulation models for the projections of spatio-temporal changes in temperature of Borneo Island based on CMIP5. Theor Appl Climatol 139(1–2):351–371

Saenphet S, Wongsawad C, Saenphet K (2001) A survey of helminths in freshwater animals from some areas in Chiang Mai. Southeast Asian J Trop Med Public Health 32:210–213

Saijuntha W, Duenngai K, Tantrawatpan C (2013) Zoonotic echinostome infections in free-grazing ducks in Thailand. Korean J Parasitol 51(6):663

Saijuntha W, Sithithaworn P, Kiatsopit N, Andrews RH, Petney TN (2019) Liver flukes: *Clonorchis* and *Opisthorchis*. In: Toledo R, Fried B (eds) Digenetic trematodes. Chapter 6, 2nd edn. Springer, Heidelberg, pp 153–199

Saijuntha W, Petney, TN, Andrews RH, Robbins R (2021a) Ticks - a large, unexplored factor in human disease transmission (this volume)

Saijuntha W, Sithithaworn P, Petney TN, Andrews RH (2021b) Biodiversity of human trematodes and their intermediate hosts in Southeast Asia (this volume)

Saintilan N, Khan NS, Ashe E, Kelleway JJ, Rogers K, Woodroffe CD, Horton BP (2020) Thresholds of mangrove survival under rapid sea level rise. Science 368(6495):1118–1121

Salmo SG III, Tibbetts IR, Duke NC (2019) Recolonization of mollusc assemblages in mangrove plantations damaged by typhoon Chan-hom in the Philippines. Estuar Coast Shelf Sci 228:106365

Sanh NH, Van Dung N, Thanh NX, Trung TN, Cooper RD (2008) Forest malaria in central Vietnam. Am J Trop Med Hyg 79(5):652–654

Sato M, Pongvongsa T, Sanguankiat S, Yoonuan T, Kobayashi J, Boupha B, Nishimoto F, Moji K, Sato MO, Waikagul J (2015) Patterns of trematode infections of *Opisthorchis viverrini* (Opisthorchiidae) and *Haplorchis taichui* (Heterophyidae) in human populations from two villages in Savannakhet Province, Lao PDR. J Helminthol 89(4):439

Schoener TW (2011) The newest synthesis: understanding the interplay of evolutionary and ecological dynamics. Science 331(6016):426–429

Schuerch M, Spencer T, Temmerman S, Kirwan ML, Wolff C, Lincke D, McOwen CJ, Pickering MD, Reef R, Vafeidis AT, Hinkel J (2018) Future response of global coastal wetlands to sea-level rise. Nature 561(7722):231–234

Sereewong C, Chaiyasaeng M, Senasri N, Chaiyos J, Tesana S (2018) Effects of physicochemical factors on development and survival of *Opisthorchis viverrini* uterine eggs. Asian Pacif J Trop Biomed 8(9):431

Shah HA, Huxley P, Elmes J, Murray KA (2019) Agricultural land-uses consistently exacerbate infectious disease risks in Southeast Asia. Nat Commun 10(1):1–3

Sheldon KS (2019) Climate change in the tropics: ecological and evolutionary responses at low latitudes. Ann Rev Ecol Evol Syst 50:303–333

Sherwood S, Webb MJ, Annan JD, Armour KC, Forster PM, Hargreaves JC, Hegerl G, Klein SA, Marvel KD, Rohling E, Watanabe M (2020) A combined assessment of Earth's climate sensitivity. Lawrence Livermore National Lab. (LLNL), Livermore, CA (United States); Jul 22, 2020

Short EE, Caminade C, Thomas BN (2017) Climate change contribution to the emergence or re-emergence of parasitic diseases. Infect Dis Res Treat 10:1178633617732296

Sibley CH (2019) Genomic analysis informs malaria evolution. Science 365(6455):752–753

Simonsen PE, Mwakitalu ME (2013) Urban lymphatic filariasis. Parasitol Res 112(1):35–44

Sinanuwong S, Takaya Y (1974) Distribution of saline soils in the Khorat basin of Thailand. Southeast Asian Stud 12:365–382

Sirivichayakul C, Radomyos PR, Praevanit R, Jojjaroen-Anant C, Wisetsing PA (2000) *Hymenolepis nana* infection in Thai children. J Med Assoc Thail 83(9):1035–1038

Sithithaworn P, Ziegler AD, Grundy-Warr C, Andrews RH, Petney TN (2012) Changes to the life cycle of liver flukes: dams, roads, and ponds. Lancet Infect Dis 12(8):588

Sodhi NS, Koh LP, Brook BW, Ng PK (2004) Southeast Asian biodiversity: an impending disaster. Trends Ecol Evol 19(12):654–660

Stephens GL, Slingo JM, Rignot E, Reager JT, Hakuba MZ, Durack PJ, Worden J, Rocca R (2020) Earth's water reservoirs in a changing climate. Proc R Soc A 476(2236):20190458

Sumarnrote A, Overgaard HJ, Marasri N, Fustec B, Thanispong K, Chareonviriyaphap T, Corbel V (2017) Status of insecticide resistance in *Anopheles* mosquitoes in Ubon Ratchathani province, Northeastern Thailand. Malaria J 16(1):299

Sumruayphol S, Apiwathnasorn C, Komalamisra N, Ruangsittichai J, Samung Y, Chavalitshewinkoon-Petmitr P (2010) Bionomic status of *Anopheles epiroticus* Linton & Harbach, a coastal malaria vector, in Rayong Province, Thailand. Southeast Asian J Trop Med Public Health 41(3):541–547

Sun Q, Miao C, Agha Kouchak A, Mallakpour I, Ji D, Duan Q (2020) Possible increased frequency of ENSO-related dry and wet conditions over some major watersheds in a warming climate. Bull Am Meteorol Soc 101(4):E409–E426

Tabakh ME, Schreiber BC, Utha-Aroon C, Coshell L, Warren JK (1998) Diagenetic origin of basal anhydrite in the cretaceous Maha Sarakham salt: Korat plateau, NE, Thailand. Sedimentology 45:579–594

Tanaka W, Wattanasiriserekul R, Tomiyama Y, Yamasita T, Phinrub W, Chamnivikaipong T, Suvarnaraksha A, Shimatani Y (2015) Influence of floodplain area on fish species richness in waterbodies of the Chao Phraya River Basin, Thailand. Open J Ecol 5(09):434

Tananchai C, Manguin S, Bangs MJ, Chareonviriyaphap T (2019) Malaria vectors and species complexes in Thailand: implications for vector control. Trends Parasitol 35(7):544–558

Tang KH (2019) Climate change in Malaysia: trends, contributors, impacts, mitigation and adaptations. Sci Total Environ 650:1858–1871

Tang Y, Bao X, Zhi Y, Wu Q, Guo Y, Yin X, Zeng L, Li J, Zhang J, He W, Liu W (2019) Overexpression of a MYB family gene, OsMYB6, increases drought and salinity stress tolerance in transgenic rice. Front Plant Sci 10:168

Tangang F, Santisirisomboon J, Juneng L, Salimun E, Chung J, Supari S, Cruz F, Ngai ST, Ngo-Duc T, Singhruck P, Narisma G (2019) Projected future changes in mean precipitation over Thailand based on multi-model regional climate simulations of CORDEX Southeast Asia. Int J Climatol 39(14):5413–5436

Tangena JA, Thammavong P, Wilson AL, Brey PT, Lindsay SW (2016) Risk and control of mosquito-borne diseases in Southeast Asian rubber plantations. Trends Parasitol 32(5):402–415

Thellmann K, Golbon R, Cotter M, Cadisch G, Asch F (2019) Assessing hydrological ecosystem services in a rubber-dominated watershed under scenarios of land use and climate change. Forests 10(2):176

Thirumalai K, DiNezio PN, Okumura Y, Deser C (2017) Extreme temperatures in Southeast Asia caused by El Niño and worsened by global warming. Nat Commun 8(1):1–8

Thompson JN (1998) Rapid evolution as an ecological process. Trends Ecol Evol 13(8):329–332

Thompson RC (2015) Neglected zoonotic helminths: *Hymenolepis nana*, *Echinococcus canadensis* and *Ancylostoma ceylanicum*. Clin Microbiol Inf 21(5):426–432

Tolentino PL, Poortinga A, Kanamaru H, Keesstra S, Maroulis J, David CP, Ritsema CJ (2016) Projected impact of climate change on hydrological regimes in the Philippines. PLoS One 11 (10):e0163941

Touati R, Haddad-Boubaker S, Ferchichi I, Messaoudi I, Ouesleti AE, Triki H, Lachiri Z, Kharrat M (2020) Comparative genomic signature representations of the emerging COVID-19 coronavirus and other coronaviruses: high identity and possible recombination between Bat and Pangolin coronaviruses. Genomics 112(6):4189–4202

Vadrevu KP, Lasko K, Giglio L, Schroeder W, Biswas S, Justice C (2019) Trends in vegetation fires in south and southeast Asian countries. Sci Rep 9(1):1–3

Van de Pol M, Jenouvrier S, Cornelissen JH, Visser ME (2017) Behavioural, ecological and evolutionary responses to extreme climatic events: challenges and directions. Phil Trans R Soc B 372:20160134

Veeravechsukij N, Namchote S, Neiber MT, Glaubrecht M, Krailas D (2018) Exploring the evolutionary potential of parasites: larval stages of pathogen digenic trematodes in their thiarid snail host *Tarebia granifera* in Thailand. Zoosyst Evol 94:425

Villafuerte MQ, Macadam I, Daron J, Katzfey J, Cinco TA, Ares ED, Jones RG (2020) Projected changes in rainfall and temperature over the Philippines from multiple dynamical downscaling models. Int J Climatol 40(3):1784–1804

Walther GR, Post E, Convey P, Menzel A, Parmesan C, Beebee TJ, Fromentin JM, Hoegh-Guldberg O, Bairlein F (2002) Ecological responses to recent climate change. Nature 416 (6879):389–395

Wang G, Cai W, Gan B, Wu L, Santoso A, Lin X, Chen Z, McPhaden MJ (2017) Continued increase of extreme El Niño frequency long after 1.5 C warming stabilization. Nat Clim Chang 7 (8):568–572

Weclawski U, Heitlinger EG, Baust T, Klar B, Petney T, San Han Y, Taraschewski H (2013) Evolutionary divergence of the swim bladder nematode *Anguillicola crassus* after colonization of a novel host, *Anguilla anguilla*. BMC Evol Biol 13(1):78

Weclawski U, Heitlinger EG, Baust T, Klar B, Petney T, Han YS, Taraschewski H (2014) Rapid evolution of *Anguillicola crassus* in Europe: species diagnostic traits are plastic and evolutionarily labile. Front Zool 11(1):74

Welcomme RL, Baird IG, Dudgeon D, Halls A, Lamberts D, Mustafa MG (2016) Fisheries of the rivers of Southeast Asia. Freshw Fish Ecol 12:363–376

Williams IN, Patricola CM (2018) Diversity of ENSO events unified by convective threshold sea surface temperature: a nonlinear ENSO index. Geophys Res Lett 45(17):9236–9244

Wongsawad C, Rojtinnakorn J, Wongsawad P, Rojanapaibul A, Marayong T, Suwattanacoupt S, Sirikanchana P, Sey O, Jadhav BV (2004) Helminths of vertebrates in Mae Sa stream, Chiang Mai, Thailand. Southeast Asian J Trop Med Public Health 35(Suppl 1):140–146

Xu C, Kohler TA, Lenton TM, Svenning JC, Scheffer M (2020) Future of the human climate niche. Proc Natl Acad Sci 117(21):11350–11355

Yang GJ, Bergquist R (2018) Potential impact of climate change on schistosomiasis: A global assessment attempt. Trop Med Infect Dis 3(4):117

Yoshida A, Doanh PN, Maruyama H (2019) *Paragonimus* and paragonimiasis in Asia: an update. Acta Trop 199:105074

Zain SN, Behnke JM, Lewis JW (2012) Helminth communities from two urban rat populations in Kuala Lumpur, Malaysia. Parasites Vectors 5(1):47

Zeng Z, Estes L, Ziegler AD, Chen A, Searchinger T, Hua F, Guan K, Jintrawet A, Wood EF (2018) Highland cropland expansion and forest loss in Southeast Asia in the twenty-first century. Nat Geosci 11(8):556–562

Zhao S, Peng C, Jiang H, Tian D, Lei X, Zhou X (2006) Land use change in Asia and the ecological consequences. Ecol Res 21(6):890–896

Zhou XN, Yang GJ, Yang K, Wang XH, Hong QB, Sun LP, Malone JB, Kristensen TK, Bergquist NR, Utzinger J (2008) Potential impact of climate change on schistosomiasis transmission in China. Am J Trop Med Hyg 78(2):188–194

Zhu F, Emile-Geay J, McKay NP, Hakim GJ, Khider D, Ault TR, Steig EJ, Dee S, Kirchner JW (2019) Climate models can correctly simulate the continuum of global-average temperature variability. Proc Natl Acad Sci 116(18):8728–8733

Zhu S, Ge F, Fan Y, Zhang L, Sielmann F, Fraedrich K, Zhi X (2020) Conspicuous temperature extremes over Southeast Asia: seasonal variations under 1.5° C and 2° C global warming. Clim Chang 160:343–360

Ziegler AD, Fox JM, Xu J (2009) The rubber juggernaut. Science 324(5930):1024–1025

Ziegler AD, Petney TN, Grundy-Warr C, Andrews RH, Baird IG, Wasson RJ, Sithithaworn P (2013) Dams and disease triggers on the lower Mekong river. PLoS Negl Trop Dis 7(6):e2166

# Chapter 10
# Praziquantel: An Efficacious Pharmaceutical Compound to Treat Human and Animal Infections Due to Trematodes and Cestodes

**Heinz Mehlhorn**

**Abstract** The anthelminthic compound praziquantel acts against flatworms (platyhelminthes) like tapeworms and bloodsucking worms (trematodes) in case they have infected humans or animals. The present chapter shows its history and ongoing activity.

**Keywords** Trematodiasis · Schistosomiasis · Tapeworm disease · Fluke control

## 10.1 History

This compound—being named as Biltricide—was developed starting in the early years after 1970 during a research project between the international pharmaceutical companies Bayer AG in Leverkusen and Merck in Darmstadt, Germany being named at first praziquantel and first marketed in the year 1980 as Biltricide®. Later followed trade names like Cesol®, Cysticide®, etc. in the case of use in humans and Tremazol®, Droncit®, Drontal®, Equimaxx®, etc. in the cases of treatment of animals. Its structural formula is shown in Fig. 10.1. Until today no significant loss of efficacy has been developed and thus praziquantel remained the drug of choice to keep humans and animals free from infections by eventually life threatening cestodes and trematodes.

H. Mehlhorn (✉)
Department of Parasitology, Heinrich Heine University, Düsseldorf, Germany
e-mail: mehlhorn@uni-dusseldorf.de



T. N. Petney et al. (eds.), *Biodiversity of Southeast Asian Parasites and Vectors causing Human Disease*, Parasitology Research Monographs 14,
https://doi.org/10.1007/978-3-030-71161-0_10

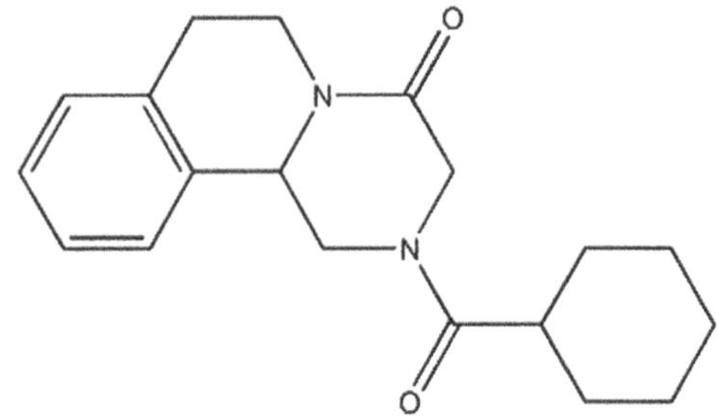


**Fig. 10.1** Structural formula of praziquantel

## 10.2 Use of Praziquantel

The compound praziquantel is orally applied and has to be given at fixed intervals and in different dosages depending on the tissue localization of the parasite in the host. It works in humans and vertebrate animals and even in aquarium fish specimens. Since more than 30 years praziquantel is constantly used in many countries (e.g. Europe, Thailand, China, Africa: see the book in this series on effects in Zanzibar: Sino-African Cooperation for Schistosomiasis Control in Zanzibar). The German company Merck KGaA offered to the WHO for 10 years more than 25 million tablets of praziquantel per year as contribution to the worldwide elimination trials of human schistosomiasis. This sum was even increased to 250 million tablets per year. They were successfully used in many countries without introduction of any signs of resistance. However, as long as the worms are not fully eliminated, reinfections remain immanent and their spreading may increase again since wild animals remain as fully active sources for new infections.

In contrast to adults very young children should not become treated with praziquantel. Before this background a cooperation was started by Merck Company (Darmstadt) in the year 2012 together with the Swiss Institute for Tropical Medicine and Public Health, Astellas Pharma Inc., TI Pharma, and the Bill and Melinda Gates Foundation to develop a suitable therapy for very young children. A success would be very helpful for these very young children and would interrupt the often noted transmission of these worms from these children to adult humans.

As it is the case in many other chemical/pharmaceutical compounds praziquantel is not completely free from side effects. The severeness of potential effects depends on the age and fitness of treated persons and thus the use of praziquantel should be done under strict control of physicians (especially in cases when also other medical compounds are used at the same time). However, praziquantel strongly helps to eliminate/avoid severe symptoms of disease in cases of infections by trematodes or cestodes (Table 10.1, Figs. 10.2 and 10.3).

**Important** The dosage of praziquantel and the period of uptake depend strictly on the advice of the caring physician, who will control the reactions of the patient. The dose and the timing of drug uptake also depend on the number of invaded *Schistosoma* species.

**Table 10.1** Attributes of praziquantel

| | |
|---|---|
| Molar mass | 312,41 g·$Mol^{-1}$ |
| Sum formula | $C_{19}H_{24}N_2O_2$ |
| Melting point | 136 °C racemic compound<br>110 °C enantiomers |
| Solubility | • Difficult, slow in water<br>• Easy in dichloromethane and ethanol |
| Toxicological data | • 2840 mg·$kg^{-1}$ rats (LD50, orally)<br>• 24,540 mg·$kg^{-1}$ mice (LD50, orally)<br>• >200 mg·$kg^{-1}$ dogs (LD50, orally) |
| Induction of potential side effects | Headache, vomiting, loss of appetite, dizziness, myalgia, fatigue, urticaria, increased body temperature |

**Fig. 10.2** Light micrograph of a couple of adult *Schistosoma mansoni* worms. The male worm transports permanently the female in a ventral fold. Both male and female have sucked the here dark appearing blood of the host

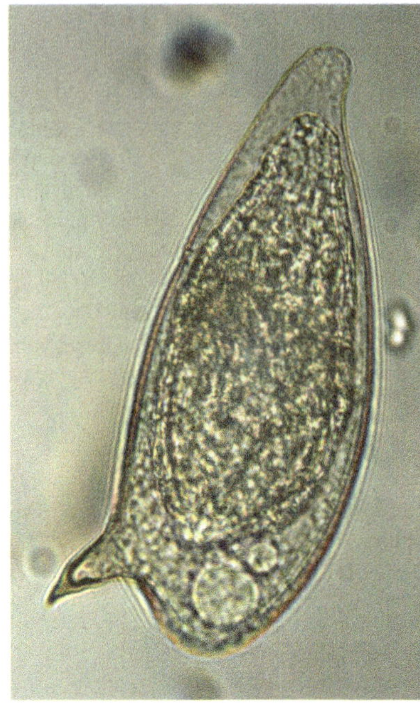

**Fig. 10.3** Light micrograph of an egg of *Schistosoma mansoni*, which is characterized by a lateral thorn and which contains already the larva (miracidium)

## Bibliography: History of important tests of efficacy[1]


Thomas H, Gönnert R (1977) The efficacy of praziquantel against cestodes in animals. Z Parasitenkd 52:117–127

Gönnert R, Andrews P (1977) Praziquantel, a new board-spectrum antischistosomal agent. Z Parasitenkd 52:129–150

Thomas H, Gönnert R (1978) The efficacy of praziquantel against experimental cysticercosis and hydatidosis. Z Parasitenkd 55:165–179

Becker B, Mehlhorn H, Andrews P, Thomas H (1980a) Scanning and transmission electron microscope studies on the efficacy of praziquantel on *Hymenolepis nana* (Cestoda) in vitro. Z Parasitenkd 61:121–133

Becker B, Mehlhorn H, Andrews P, Thomas H, Eckert J (1980b) Light and electron microscopic studies on the effect of praziquantel on *Schistosoma mansoni*, *Dicrocoelium dendriticum*, and *Fasciola hepatica* (Trematoda) in vitro. Z Parasitenkd 63:113–128

Becker B, Mehlhorn H, Andrews P, Thomas H (1981) Ultrastructural investigations on the effect of praziquantel on the tegument of five species of cestodes. Z Parasitenkd 64:257–269

Mehlhorn H, Becker B, Andrews P, Thomas H, Frenkel JK (1981) In vivo and in vitro experiments on the effects of praziquantel on *Schistosoma mansoni*. A light and electron microscopic study. Arzneimittelforschung 31:544–554


[1]Literature on early detections on the efficacy of praziquantel in chronological order showing some of the huge number of initial cooperations of international groups with scientists of Bayer Company, Leverkusen, Germany.

Cooperative Group of Treatment of Fasciolopsiasis with Praziquantel (1981) A comparison of the curative effects of praziquantel, bithional and areca pill in the treatment of fasciolopsiasis. Jiangsu Med J 7:5–7

Mehlhorn H, Frenkel JK, Andrews P, Thomas H (1982) Light and electron microscopic studies on *Schistosoma mansoni* granulomas of mouse livers following treatment with praziquantel. Tropenmed Parasitol 33:229–239

Mehlhorn H, Kojima S, Rim HJ, Ruenwongsa P, Andrews P, Thomas H, Bunnag B (1983) Ultrastructural investigations on the effects of praziquantel on human trematodes from Asia: *Clonorchis sinensis*, *Metagonimus yokogawai*, *Opisthorchis viverrini*, *Paragonimus westermani* and *Schistosoma japonicum*. Arzneimittelforschung 33:91–98

Abo-Shady AF, Abdel-Magied SA, Hegazi MM (1983) Praziquantel in treatment of heterophyiasis (a preliminary study). J Egypt Soc Parasitol 13:629–630

Sirisinha S, Puengtomwatanakul S, Sobhon P, Saitongdee P, Wongpayabal P, Mitranonde V, Radomyos P, Bunnag D, Harinasuta T (1984) Alterations of the surface tegument of *Opisthorchis viverrini* exposed to praziquantel in vitro and in vivo. Southeast Asian J Trop Med Public Health 15:95–103

Wahn V, Mehlhorn H (1984) Four kinds of parasites in an 8-year-old boy. Therapeutic effects of praziquantel against *Fasciola hepatica*. Dtsch Med Wochenschr 109:1486–1488

Schmahl G, Mehlhorn H (1985) Treatment of fish parasites. 1. Praziquantel effective against Monogenea (*Dactylogyrus vastator*, *Dactylogyrus extensus*, *Diplozoon paradoxum*). Z Parasitenkd 71:727–737

Taraschewski H, Mehlhorn H, Bunnag D, Andrews P, Thomas H (1986) Effects of praziquantel on human intestinal flukes (*Fasciolopsis buski* and *Heterophyes heterophyes*). Zentralbl Bakteriol Mikrobiol Hyg A 262:542–550

Staudt U, Schmahl G, Blaschke G, Mehlhorn H (1992) Light and scanning electron microscopy studies on the effects of the enantiomers of praziquantel and its main metabolite on *Schistosoma mansoni* in vitro. Parasitol Res 78:392–397

Mehlhorn H, Schmahl G, Frese M, Mevissen I, Harder A, Krieger K (2005) Effects of a combinations of emodepside and praziquantel on parasites of reptiles and rodents. Parasitol Res 97 (Suppl 1):S65–S69

Mehlhorn H (ed) (2016) Encyclopedia of parasitology, 4th edn. Springer, Berlin, Heidelberg

The manufacturer's authorised representative in the EU is Springer
Nature Customer Service Centre GmbH, Europaplatz 3, 69115 Heidelberg,
Germany. If you have any concerns regarding our products, please
contact ProductSafety@springernature.com

Printed and bound by CPI Group (UK) Ltd, Croydon, CR0 4YY
28/09/2026
02224091-0005